ENGLISH LITERATURE AND THE WIDER WORLD

General Editor: Michael Cotsell

Volume 4
1876–1918

The Ends of the Earth

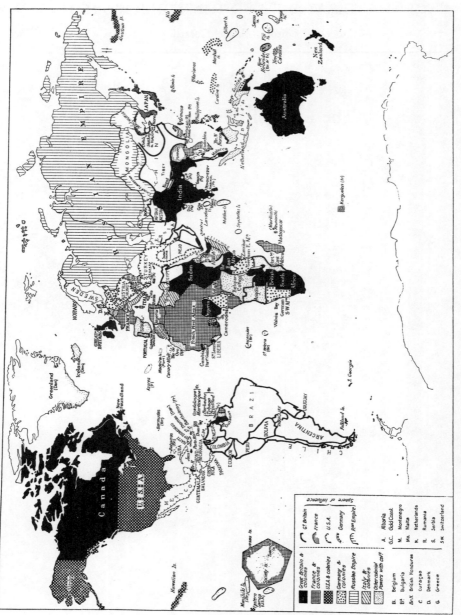

"The First Division of the World, 1876–1913," from Alexander Rado and M. Rajchman, *The Atlas of To-Day and To-Morrow* (1938).

£3

April 10, 1995
London
D. Finkelstein

ENGLISH LITERATURE AND THE WIDER WORLD

General Editor: Michael Cotsell

Volume 4

1876–1918

The Ends of the Earth

edited by

SIMON GATRELL

The Ashfield Press

London – Atlantic Highlands, NJ

First published in 1992 by The Ashfield Press Ltd.,
3 Henrietta Street, Covent Garden, London WC2E 8LU, and
Atlantic Highlands, New Jersey 07716

Library of Congress Cataloging-in-Publication Data
(Revised for vol. 4)

English literature and the wider world.

Includes bibliographies and indexes.
Contents: v. 1. All before them, 1660–1780 / edited
by John McVeagh — v. 3. Creditable warriors, 1830–
1876 / edited by Michael Cotsell — v. 4. The ends of
the earth, 1876–1918 / edited by Simon Gatrell.
1. English literature—Foreign influences.
2. Travel in literature. 3. Exoticism in literature.
4. Authors, English—Journeys. I. Cotsell, Michael.
II. Gatrell, Simon. III. McVeagh, John.
PR125.E54   1990   820'9.'008   88–7489
ISBN 0–948660–11–2 (v. 4)

A catalog record is available for this book from the British Library.

Printed in the United States of America

For Tony

Oh, East is East, and West is West, and never the twain shall meet,
Till Earth and Sky stand presently at God's great judgement Seat;
But there is neither East nor West, Border, nor Breed, nor Birth,
When two strong men stand face to face, though they come from the ends
  of the earth!

Kipling, "The Ballad of East and West," 1889

The old river in its broad reach rested unruffled at the decline of day, after ages of good service done to the race that peopled its banks, spread out in the tranquil dignity of a waterway leading to the uttermost ends of the earth.

Conrad, *Heart of Darkness*, Chapter 1

"That is why," the girl continued, "I am going to Glasgow—to take my mother away from there." She added: "To the ends of the earth."

Ford, *The Good Soldier*, Part 4, Chapter 2

CONTENTS

CONTENTS

LIST OF ILLUSTRATIONS

ILLUSTRATIONS

GENERAL EDITOR'S PREFACE

The expansion of Europe is arguably the most momentous development of modern history. Britain has played a major role in that expansion. As tourists, travelers, and explorers; as traders and colonists; as warriors and imperialists, the British have been an expanding people. The idea has been familiar for three hundred years and is a recognized thematic for the historian. But in literary studies the overall significance of British expansion for the understanding of what is called English literature has not been fully addressed. Recent critics and scholars—Martin Green, Edward Said, Gayatri Spivak—have begun to point to this fact and to speculate about what is a surprising, even disturbing omission.

Literary history, it seems, has been developed to reinforce an ethos of homogeneous and high-minded nationhood. We may think that two things have been concealed in the traditional account: the relation of the imagination to power and the demand of the national imagination for something more than the national life has provided. Comparative literature has indeed studied analogous developments in cultures and traced influences of ideas and forms, but has not explored the manner in which English literature has been shaped by the undeniable experience of what we are here calling "expansion," and the extent to which it has sought to supplement the domestic experience.

The series *English Literature and the Wider World* seeks to rectify these omissions. The four volumes of the series, dealing with the period 1660–1918, demonstrate that the response to the geography and peoples of the non-English world has always been a major part of the story of the British literary imagination. They show that few literary genres or writers of significance can be properly understood without consideration of this perspective.

Each volume contains a substantial editor's introduction, which discusses the political events of the period, conditions of travel, the literary response to the wider world, and the response to the particular regions of the world. The introductions provide the context for a dozen or so essays by contributing scholars which read the important authors in each period whose work draws significantly on the non-English world; this turns out to be the majority of important English writers. A guide to authors' travels and to further reading completes each volume.

The essays represent a variety of critical approaches, and it is hoped that they will provide stimulus to further interpretations. Of necessity, the essays concentrate on major figures. The unwished-for effect is that minor writers and travel writers of great interest go relatively neglected outside of their treatment in the introductions. But the series will encourage further studies of relations between the canon and writers on its geographical and other margins.

The series looks at how the British have imagined the world. It is in a sense an account of both an experience and an illusion—that of British centrality—and it thus runs the risk of duplicating anglocentrism. We can only hope that we have proceeded with some awareness of this limitation. The term "English literature" has been retained, though it refers also to Scots and Welsh literature: we have tried to recognize that the British margins may well experience different relations with the world beyond Britain. Ireland has been treated as part of the wider world, although for the period covered by these volumes it was, it one way or another, politically attached to Britain. The reader who is unhappy with this choice might like to consider the alternative. British interrelations with Ireland have been and are complex: the inclusion of writers of Irish birth or residence in this volume is not an attempt to appropriate them for a British tradition.

The series has been supported by a generous subvention from the University of Delaware, and in this connection I would like to thank Provost Leon Campbell, Dean Helen Gouldner, and Professor Jay Halio.

The series is gratefully dedicated to Mario Pazzaglini.

MICHAEL COTSELL
University of Delaware

ACKNOWLEDGMENTS

I should like to thank Emily Hipchen and Michael Cotsell for reading and discussing with me parts of my share of this volume; the whole series owes a great debt to Michael's care and enthusiasm. I should like also to thank Gil Lawson, Tim Richardson, Alice Kinman, and Glenn Harris for the assistance that they have given in the preparation of this volume, and the Department of English at the University of Georgia for making their help available to me.

INTRODUCTION

1. Political Developments

The plot of the drama outlined in this section of the Introduction is one of dual intertwining threads, Europe and Empire; its action reveals the growth toward conflict of dynastic houses complexly related by marriage. The stakes for these imperial powers are high (at first the political and economic leadership of Europe, and later nothing less, it seems to the participants, than world domination). There is hubris everywhere, and in pursuit of ambitions there are skirmishes at remote ends of the earth throughout the action; it seems a classic history play until suddenly the war of words and scuffles turns frighteningly universal. After four years, three of the dynasties fall (at last, the moment Thomas Hardy was looking for), but many millions die. Perhaps then we might consider it a tragedy and not a history play—the fall of kings, revolutions in the wheel of fortune, reliance upon divine authority; but if so it is a modernist tragedy; at the end of the first (and only) performance there was no cathartic relief, since there was no audience to experience it; all were participants. Nor was there amongst the victorious anyone to pronounce the required words of recognition and reassurance—no one clear-sighted enough to see beyond victory and defeat and strong enough to enforce the vision. Ford Madox Ford had the phrase to describe this drama: it is indeed the saddest story.

The story begins with three events that occurred between 1875 and 1878, events that can be thought of as highly significant in the development of relations between England and the rest of the world: in 1875, Benjamin Disraeli, financed by Lionel and Nathan Rothschild, bought the Khedive of Egypt's 40 percent holding in the Suez Canal; in the following year Victoria was proclaimed Empress of India; and in 1878 the Congress of Berlin brought relative calm to the Balkans, where Russia had decisively defeated the Ottoman Empire, and thirty years of peace to Europe.

The construction of the Suez Canal in 1869 fundamentally altered England's foreign and colonial policy by providing a quicker and less dangerous route to India and the East; the purchase of the Khedive's shares brought England direct involvement in the politics of Egypt and the Eastern Mediterranean.

1

For forty years attitudes to England's colonial possessions had been formed along conservative financial lines; it had been conventional wisdom to suggest that unless the colonies could be self-supporting they were an economic liability—in Disraeli's phrase "millstones round our necks." Awarding self-government in various degrees to the white-dominated settlement colonies in Canada (1867), Australia (various states from 1855 to 1893, the Australian Commonwealth in 1901), New Zealand (the earliest of all in 1841), the Cape Province (1872), and Natal (1893) was one manifestation of this economic prudence. But when Disraeli at last became prime minister in 1874, he was already sensitive to a new mood in the country, more receptive than heretofore to ideas of Imperial glory. The short-lived second French Empire had recently been destroyed; the German Empire had just been proclaimed; dynastic ambitions were in the air in central Europe. Disraeli understood the nascent anxiety in England that new economic and political rivals were emerging, and his proclamation of Victoria as Empress of India made a powerful symbolic assertion of England's long-standing but hitherto untrumpeted Imperial grandeur.

In Europe the eruption of Bismarck's Prussia to prominence through military defeats of the Austrian Empire (1865) and France (1870), and the consequent consolidation and expansion of a German Empire under the Prussian monarch, altered completely the diplomatic landscape. It also inaugurated the complex ballet of alliances and ententes that was performed across the face of Europe until the final transformation scene in 1914–18. There remained only one area of Europe relatively unprotected from expansionist ambitions, the European provinces of the Ottoman Empire in the Balkans. The Crimean War (1854–56) and its mismanaged peace settlement had for a while propped up Moslem rule in this part of Europe, but in 1877 war again broke out between Russia and Turkey. It seemed that Russian forces would be in Constantinople by July, but the unexpected resistance of the Turkish fortress at Plevna, in the direct line of their march, held them up for four months, by which time England had again decided to block Russian imperial ambitions, in particular for the establishment of a port on the Mediterranean. Though the intervention was made reluctantly on the part of the English government, the mood of at least some of the people may be gauged by the famous lines from a contemporary music-hall song:

We don't want to fight, but by jingo if we do
We've got the ships, we've got the men, we've got the money too!

For six months an English fleet in the Sea of Marmora stood between the Russians and Constantinople, until in the summer of 1878 the Congress of Berlin was convened, at which the Great Powers, orchestrated by Bismarck, settled the war by returning most of the conquered provinces to the

Ottoman Empire. England was granted a protectorate over Cyprus, strategically important in relation to the Suez Canal, while the joint financial control of Egypt was arranged with France in return for giving her a free hand in Tunis. But more important to England than the actual terms of the settlement was the fact that the Congress of Berlin became the inaugural event of the great European Imperial tournament that was to help shape politics for the next forty years, in England no less than in Germany, Russia, France, Italy, Spain, or (as a late-entering visiting player) the United States. The chief victims of this blood sport were the decaying Ottoman Empire (removed from Europe by 1914, and reduced to modern Turkey by 1918), China (the partition of which seemed imminent every year from 1898 onwards, but was never accomplished), Africa in general (the division of which amongst the powers was complete well before 1914), and various small islands in the Pacific. South America, though tempting no doubt, was put out of bounds (by the Monroe Doctrine) to all except the United States for anything but commercial exploitation, of which there was plenty.

From 1876 to the conclusion of the Great War in 1918, English foreign politics were dominated by this entanglement of European and Imperial concerns; the main focus of colonizing action for all the powers was to the immediate south, in that ruthless exercise in arbitrary partition familiarly known as the scramble for Africa (the undignified overtones in the phrase are vividly appropriate). England's part in the fun began at the north and south of the continent and rapidly spread inwards to the center. Her political and (perhaps more powerfully effective) financial interest in the security of Egypt led to the military suppression of a nationalist rising under Arabi Pasha in 1882, and the careful control of all subsequent nationalist movements—a task managed by Evelyn Baring, later Lord Cromer, who along with Alfred Milner (who worked with Baring for a while), was the most effective and enduring of Imperial agents. A series of bargains with France (which also had a strong financial stake in Egypt) led to virtual English hegemony, though Egypt remained nominally a part of the Ottoman Empire—as did the Sudan to the south, occupied by the Egyptians in 1820.

In 1881 under the Mahdi, the inhabitants of the Sudan rebelled against Egyptian control. By this time William Ewart Gladstone was again prime minister; he retained essentially the mid-Victorian view of colonial expansion, and ordered the evacuation of the Sudan, rather shortsightedly sending—at the beginning of 1884—an evangelical Christian, General George Gordon, to supervise the withdrawal. Gordon's temperament and conscience would not allow him to retreat and leave the country under the control of what he thought of as bloodthirsty and fanatical Muslim rebels. He asked for reinforcements and planned a counterattack; Gladstone was

3

slow to authorize either, and gradually, as Gordon was cut off in Khartoum, religious and patriotic opinion in England mobilized against the government's vacillation. Finally, in the autumn of the year, Garnet Wolsely was sent to Egypt and a large force began cautious operations down the Nile, which resulted a few months later in their arrival before Khartoum two days too late to save Gordon. This incident was another important marker in the rise of jingoism in the English public. Gordon was widely regarded as a martyr, though he had deliberately disobeyed the orders of his government; and many people, amongst them the queen, saw Gladstone as the instrument of his martyrdom. He had to resign.

In 1884 the process of arbitrary division picked up pace; Germany made good spectacular claims in south-west and eastern Africa and the Cameroons, and in the same year the independent status of the Congolese region claimed by King Leopold II of the Belgians was formally recognized as a "free state" in the "interests of humanity," particularly as a bar to Mahdist progress in Central Africa. In fact, the Mahdist state in the Sudan remained a sore point in the consciousness of the new Imperialists until, at the battle of Omdurman in 1898, it was suppressed by Kitchener—as far as the jingos were concerned, a satisfying act of revenge for the murder of Gordon. Wilfrid Blunt in his published diary (*My Diaries*, 2 vols., 1919) wrote of this victory: "The whole country, if one may judge by the Press, has gone mad with the lust of fighting glory, and there is no moral sense left in England to which to appeal" (1.296). A year earlier Blunt had recorded a discussion in which each participant gave his idea of Heaven; the poet-laureate Alfred Austin's idea "was to sit . . . in a garden, and while he sat to receive constant telegrams announcing alternately a British victory by sea and a British victory by land" (1.280).

In South Africa in the mid 1880s massive reserves of gold were discovered in the Transvaal, an event which was to have long-term consequences. In 1877, when Disraeli was in power, the Boer Republic of the Transvaal had been annexed by the administrator of Natal, Theophilus Shepstone, since it appeared to be unable or unwilling to protect itself against attacks by neighboring African nations. In 1880–81 the Boers revolted and defeated an English force at Majuba Hill in Natal; Gladstone was content to negotiate a settlement that returned virtual independence to the Transvaal. The subsequent discovery of gold made it inevitable that the conflict between the Boers and the English should be renewed. The Boers needed English capital to develop the mines, but they were anxious to prevent either the financiers or the Cape administrators (in particular Cecil Rhodes, who was both very rich and prime minister of the Cape) from gaining political influence within the state. They began building a railway to the Indian Ocean through Portuguese Mozambique; they imposed heavy taxation on the mines and the

4

"The English Pursuit of De Wet" (1901), by Rudyard Kipling (*Illustrated London News* 118.375).

gold mined; and they refused to give foreign residents the right to vote.

Ultimately the last of these grievances was to provide the excuse for hostilities, though many in England and elsewhere recognized that the true struggle was over political and economic control of South Africa. War was engineered in 1899; it took considerably longer to defeat the Boers than had been anticipated. The inefficiency of the English army, and in particular the incompetence of much of its leadership, was made evident. The war (particularly what was reported—often inaccurately—about the camps in which many Boer families were placed for their safety, and which came to be called concentration camps) made England for a while unpopular throughout Europe, and it was only the unquestioned supremacy of the English navy that prevented concerted action by the other great powers. Ultimately it took more than three years to defeat the Boers.

The war, coming at the end of the century and the end of Victoria's reign, proved to be of as much symbolic as political importance. It marked the end of the Victorian political and social self-confidences, the final end to laissez-faire colonialism, the end of warfare that engaged only a minority of the people. Because it provided vivid evidence not only of England's naval power, but also of possible weaknesses in that power, the war marked the beginning of open preparations for pan-European, even global, conflict as Germany, France, Russia, and Japan added to universal conscription huge programs of naval development, accelerated (especially in Germany) after the launching by England in 1906 of the all-big-gun *Dreadnought*, a battleship that rendered all existing navies obsolete. Blunt was friendly with the young Tory Imperialist member of parliament (MP) George Wyndham, and reports him as saying "that it is now simply a triangular battle between the Anglo-Saxon race, the German race, and the Russian, which shall have the hegemony of the whole world. . . . George and the young Imperialists are going in for England's overlordship and they won't stand half-measures or economy in pushing it on" (1.322). Concurrently and almost paradoxically, it marked in England the beginning of a fresh social optimism, the tone for which was set in part by the new king, Edward VII, and in part by the increased influence of socialist ideals within the Liberal party, which was in power for the majority of the next fifteen years.

Ireland, though, had been a political problem for the English throughout Victoria's reign, and remains one still. The essential structure of the narrative is well known. The special psychological and emotional difficulties arose primarily because of the physical closeness of Ireland to England—it was easier to allow Canada or the Australian states to govern themselves; also many of the owners of land in Ireland were politically influential in England. There is no doubt that geographically Ireland is part of the British Isles, but to Irish and English alike there was equally no doubt that there were sufficient cultural and religious distinctions between English and Irish to make the situation one of colonial rule. Indeed, economically and fiscally Ireland was treated like a colony.

Within Ireland there were successive but abortive attempts at insurrection to end English rule, combined from the 1880s with a more effective campaign to return the ownership of the land to the people. There was a broadly satisfactory settlement of the Irish Land Question passed through Parliament in 1903. Also at Westminster, under pressure from the group of Irish MPs effectively organized by Charles Stewart Parnell until his death in 1891, which often during these years held the balance of power between Conservative and Liberal, there were five attempts to pass bills that would grant Home Rule to Ireland; all were defeated. Even if they passed the Commons, the huge inbuilt Conservative majority in the Lords guaranteed failure. Anger at the intransigence of the Lords in this matter ultimately led in 1910 to a constitutional crisis in which the Liberals threatened to create enough peers to ensure a majority for the passage of Home Rule (and of social welfare) legislation. As Home Rule became more and more of a probability, there was aggressive resistance to the idea in predominately Protestant Ulster, organized under Edward Carson and the slogan "Ulster will fight, and Ulster will be right," until, in 1914, civil war seemed a real possibility.

Beyond Europe, the United States had by 1901 also become an imperial power, reaching across the Pacific to annex Midway, Wake Island, Hawaii, and finally—after the 1898 war with Spain—Guam and the Philippines (upon which conquest Kipling urged Americans to take up the "White Man's Burden"). In the Caribbean the U.S. acquired Puerto Rico and protectorates over Cuba, Haiti, the Dominican Republic, and, during the first decade of the new century, Panama and Nicaragua. It soon became clear that their Pacific competitor would be Japan rather than Russia; in 1904–1905 the Japanese dealt a sequence of humiliating and crushing defeats to the Russian army and navy on the Manchurian mainland. Blunt commented, alluding to the conduct of the Boer War: "How stupid our English Generals must feel when they see the strongest fortress in the world taken, fort after fort, by storm by the Japanese" (2.113). A little later, after the Japanese capture of Mukden, he noted hopefully: "it is the first great victory of the East over the West since the Ottoman conquests of the sixteenth and seventeenth centuries" (2.117).

As the new century began in Europe there was much diplomatic maneuvering, but not until 1908 was there "the first premonitory thunderclap of what was presently to prove the great storm which, beginning with the aggression of Italy on Turkey, was to involve Eastern Europe in a series of wars, and eventually the whole Western world in the overwhelming catastrophe of 1914" (*My Diaries* 2.20). In April of that year, Italy, anxious to

7

catch up with the other powers in the colonial race, invaded the Ottoman province of Tripoli; the invasion was hardly successful, but it began a concerted attempt to dismember the Turkish Empire. In one week in October, Austria-Hungary annexed Bosnia and Herzegovina, Bulgaria declared itself independent, Crete annexed itself to Greece, and Albania declared itself independent. There were democratic revolutions in Turkey and military defeats, until a tentative compromise was reached in 1909. At any time almost, had Germany desired, a pan-European conflict might have been engineered over these Balkan issues, which were primarily fuelled by the rivalry between Austria and Russia for influence in the area. But the final defeat of the Turks in the Balkan war of 1912–13—removing them altogether as a power in Europe—seemed to be the last act of this drawn-out struggle, that is, until a year later when the local problems were subsumed in the greater conflict.

In North Africa, too, there were crises that might have been made the flashpoint of large-scale fighting; the Moroccan conflict of 1905–1906 raised the serious possibility of war between France and Germany, a war that would almost certainly have involved England on France's side. The incident prompted, for the first time, informed and active English politicians to propose that Germany was maneuvering for European hegemony. The second Moroccan crisis of 1911 was still more dangerous; England for a time strongly objected to the acquisition by Germany of an Atlantic port on the coast of Morocco at Agadir, and some preparations for war were actually taken.

Further south the business of acquiring Africa continued, as Frederick Lugard extended Imperial territory in Nigeria beginning in 1903. However, as the Empire upon which the sun never set spread still more widely, there came also signs of strain. There was concern in reconstructed federated South Africa about a possible reintroduction of slavery, and there had been a sharply squashed rising of the Zulus in 1906. In India there were Nationalist riots in Rajasthan and in Bengal (particularly in 1908), and it began to be possible, for those who wished, to anticipate a day when the British Raj would crumble as others had done.

Until the very last, many in England did not believe that their country would become involved in the war that began at the end of July 1914 in the Balkans as part of the ongoing conflict between Austria-Hungary and Russia for domination there and that spread rapidly through the network of European alliances to include Germany and France. The English army was small, and though some urged support for France, it did not seem that an English contribution by land could be significant; however Herbert Asquith, Sir Edward Grey, and Winston Churchill thought differently, and with the German invasion of Belgium as the pretext, war was declared on

Germany and her allies on August 5, 1914. Imperial expansion of all the conflicting powers meant that the theater of the war would be worldwide (the British Empire covered a quarter of the land surface of the globe, and accounted for a quarter of its population), though the primary focus of attention in England was on the western front entrenched across Flanders and Picardy. Though none of the warring nations realized it, the state of development of armaments and of the tactics of warfare ensured that millions of men would fight and die, and that the whole populations of the combatants would be directly affected, both economically and through the loss of relatives or close friends. For three years there was much activity and slaughter, none of it decisive and much of it quite ineffective, and it was the intervention (brought about by unrestricted German submarine warfare) of American troops on the Anglo-French side in 1917 that provided the thrust that broke the European stalemate. When this direct involvement of the United States in European affairs is considered together with the Bolshevik Revolution in Russia in the same year, it is evident that there was thus inaugurated a new phase of world politics, a phase that may only now be drawing to a close.

2. Travel and Travel Writing

It is part of the record of technological progress with which we are nowadays so much at home that almost everywhere during this period the speed, facility, and comfort of transportation increased. The traveler could reach more places with less effort than in previous years and (to make a distinction the validity of which has been debated recently) the tourist could reach familiar places in less time and with less anxiety. The Introduction to Karl Baedeker's first English edition of his handbook to Greece (1889) begins:

> A journey to Greece no longer ranks with those exceptional favours of fortune which fall to the lot of but few individuals. Athens, thanks to modern railways and steamers, has been brought within four days of London. . . . The number of travellers who, after exploring Italy and Sicily, turn their steps towards the classic shores of Hellas, the earliest home of the beautiful, will therefore doubtless constantly increase. (*Greece*, xi)

One thinks, for instance, of the elderly Miss Alans in E.M. Forster's *A Room with a View*, bravely and insouciantly contemplating travel not only to Greece, but also to Constantinople, armed with stocks of digestive bread.

In his essay "The Regrets of a Veteran Traveller," collected in *Memories and Thoughts* (1906), Frederic Harrison wrote in 1887:

9

Railways, telegraphs and circular tours in twenty days have opened to the million the wonders of foreign parts. But have they not sown broadcast disfigurement, vulgarity, stupidity, demoralisation? Europe is changed indeed since the unprogressive forties. (219)

Harrison went on to suggest that the small European wars of the past thirty years had driven travelers of different nationalities, who formerly would have conversed amicably together in the coach or the single modest inn of a small town, to stick to their own and to shun the foreigner; so that instead of traveling in order to understand something of the ways and ideas of other peoples, each nation now established its own hotel, and imported its own way of life.

Morally, we Britons plant the British flag on every peak and pass; and wherever the Union Jack floats there we place the cardinal British institutions—tea, tubs, sanitary appliances, lawn tennis, and churches. . . . In things spiritual and things temporal alike our modern mania is to carry with us our own life, instead of accepting that which we find on the spot. (226, 233)

Eleven years later Harrison wrote in praise of the Italian Riviera east of Genoa in "The Riviera di Levante," lamenting in similar terms the replacement of the characteristic environment of the French Mediterranean coast with international cosmopolitanism: "Metropole Caravanserais of the latest pattern." What is particularly interesting about this essay as collected in *Memories and Thoughts* is that Harrison added a postscript in 1906:

Alas! alas! this corner, too, of old Italy is going the way of all else that was lovely, sacred, and historic in Europe. American Grand Hotels, Monte Carlo villas, Parisian boulevards have already invaded this peaceful retreat of our old age; and I am told that my own praises of it have helped to swell the incursion of our Northern barbarians. And now that new disease, the pestilent *motoritis*, has begun to make the Riviera di Levante as foul, as noisy, and as dusty as is the Riviera di Ponente at the height of its orgies. (244)

There are three points to derive from this Afterword: the habitual, paradoxical fate of effective travel writers—that they inevitably contribute to the degradation by feet and bodies of what they have so vividly or lovingly evoked; the habitual nineteenth-century thought that barbarian invasions always come from the north; and the habitual lament over the arrival of the motorcar as an instrument of tourism.

10

Much of the original impetus behind the changes noted and mourned by Harrison was provided by Thomas Cook—though of course he was responding to the mid-century growth in the European railway system and the mid-Victorian growth in affluence of the middle and lower-middle classes. Something has already been said about this almost legendary entrepreneur in the introduction to Volume III of this series. By 1876, Cook's had established throughout Europe the patterns of modern tourist travel made easy and secure by guarantees from the company; they had successfully organized the first round-the-world tour (in 1872, the same year that Jules Verne was publishing *Around the World in Eighty Days*); and they had established the feasibility of safe and economical group travel in Palestine. The present period was to see their greatest public successes: the most notorious coup was the tour they arranged for Kaiser Wilhelm II to the Holy Land (*Punch* called the members of the tour "Cook's Crusaders"), but there were a number of other highly significant enterprises.

Cook's developed travel on the Nile almost single-handedly, building hotels and a hospital at Luxor, and being made responsible by the Khedive in 1880 for all the passenger steamers on the river. When Arabi Pasha's rising stopped tourism, and the metropolitan authority decided to destroy the rising, the British army's leaders traveled out to Egypt on Cook's tickets and the Cook's Nile steamers carried dead and wounded from the decisive slaughter of Tel-el-Kebir. When the Mahdi rebelled in the Sudan, General Gordon's trip to Khartoum with instructions to oversee the evacuation of the country was arranged by Cook's, as was the whole of the subsequent abortive expedition to rescue Gordon. Later the Nile fleet was rebuilt and Cook's ruled the river until the Second World War. Again, the company facilitated and organized travel between India and England, conducting through India many of those "globe-trotters" that Kipling (speaking for Anglo-India) found so obnoxious, and escorting Indian princes through Europe. A further development was the assumption of the arrangements for Indian pilgrims visiting Mecca.

As the century turned ominously for the English with the death of Victoria and an uneasy state bordering on humiliation in South Africa, the motorcar began to be seen more and more frequently on English roads. In England a strict speed limit was imposed, and there were many speed traps; on the Continent, however, there were no speed limits, and thus another motive for traveling abroad was born. As the motoring organizations, particularly the Automobile Association (with eight-thousand members by 1909), became involved in the complicated business of shipping the cars across the English Channel, and seeing them through French customs, so it became relatively easy to drive to Rome or Biarritz. Act 2 of George Bernard Shaw's *Man and Superman* (1903) is played with a motorcar on stage, and

"Arrival of the Post Boat at Assouan" (1884), Melton Prior (*Illustrated London News* 85.448).

just such a transcontinental trip is planned. In 1906, Edward VII when visiting Vesuvius demanded of the Cook's agent in Naples that he should be driven to their funicular railway in a motorcar—which, after some trouble, Cook's procured. By 1912, James Scully was able to suggest that touring automobiles were sufficiently numerous to restore life to networks of roads that had been decaying since the development of the railway systems of Europe: "The new motor can claim, as a set-off to its noise, smell and other unpleasantness, the credit of having revived many an old and moribund country road. It is reviving the Mont Cenis; for this pass is an easy one for the motorist."[1]

There were, however, to 1918 and beyond, still places in the other continents where Cook's arm did not reach and the motorcar could not penetrate, places where a white representative of pan-European culture could expect to be the first of his or her kind to set foot; and there were consistently people anxious to explore them and to write about their experiences. Indeed, as the habit of traveling abroad on a regular basis percolated further and further down the socioeconomic strata in England, so the corresponding springs of travel writing issuing from the publishing houses grew more and more in volume of flow. At one extreme there was the guide book, with maps and facts and pseudo-authoritative opinions, forming in the crudest way possible the attitudes of the reader; at the other extreme there was the work of the creative imagination taking as part of its focus the experience of an area outside England (the material for almost all the essays in this volume).

This period saw the emergence of Baedeker's guides as rivals to John Murray's. Where Claude in Arthur Hugh Clough's *Amours de Voyage* peers round Rome with Murray in hand, Lucy Honeychurch in Forster's *A Room with a View* has Baedeker to help her in Florence. There was not much difference between them, though Baedeker tended to be more discursive in a scholarly manner; perhaps the choice of one or the other was a matter of fashion.

Such guide books are still relatively familiar, but guides to the dominions and colonies during this period were rather different, and have almost entirely been superseded. The title of one such handbook indicates the nature of the difference: *The Guide to South Africa for the Use of Tourists, Sportsmen, Invalids and Settlers* (ed. A. Samler Brown and G. Gordon Brown). In the 1900–1901 eighth edition of this guide there are hints that the Boer War might affect certain aspects of life in South Africa, but throughout it is regarded as a merely local disturbance of the real businesses of exploitation of the mineral and natural and human resources of the country. There are provided exhaustive details concerning all matters that any of the suggested users of the volume might need, from eleven pages on South

Africa as a health resort, to a sharp little note on the incidence of the livestock disease scab and the penalties for not dipping one's sheep: "The compulsory Scab Act of 1894 is now law" (172A); or from five pages on irrigation to twenty-five pages on the game of South Africa. There are twenty-seven pages summarizing the history of South Africa, but there is (unsurprisingly) almost no instruction concerning the black inhabitants of the region. After the factual Introduction there follow descriptive itineraries in a form familiar to Murrayolaters, but with a very different content. The note is one of amazing and rapid progress amidst desert land, of fertility of capital and investment, of size and modernity, often stressing the achievements independent of metropolitan assistance.

Another feature of guide books seems to be that they invite physical interaction with the user; it is common to find in second-hand bookshops copies of guide books with vivid marginal notations, as is the case with the particular copy of the South African one referred to above. It is instructive to imagine Arthur Pegler, the first owner of this volume, traveling out from Southampton on the same Union-Castle steamer as Kipling in 1900, undeterred by the war from his plan of settling near Pietermaritzburg. From one of the marginal notes it seems he had some idea of trying to grow coffee, and his evident interest in the health-promoting qualities of the climate of upland Natal suggests that he, or someone traveling with him, had a respiratory complaint.

From the base-camp of the guide book, with its honestly displayed purpose, there is a long outward path over different qualities of travel writing. It is probably no coincidence that most of the finest explorer-writers during the period were women: Mary Kingsley in West Africa, Gertrude Bell in Arabia, or Isabella Bird (who published more than half of her work under her married name of Isabella Bishop) all over the world. Whether or not they admitted it to themselves, they were more than topographical explorers; they were rebels against the dominant conventions of late Victorian society concerning women, and this added several fresh perspectives to their books. As explorers they took greater pains to ensure the accuracy of their observations, and they were mostly conscious of interpreting what they saw from a distinctively non-male point of view. As writers they accepted most of the established conventions of the male-dominated genre, but added to them, consciously or unconsciously, an extra level of intensity.

Isabella Bishop was perhaps the most influential of these women adventurers. A list of her major books gives some idea of the scope of her traveling: *Six Months in the Sandwich Islands* (1875); *A Lady's Life in the Rocky Mountains* (1879); *Unbeaten Tracks in Japan* (1880); *The Golden Chersonese* (on the Malay peninsula) (1883); *Journeys in Persia and Kurdis-*

14

tan (1891); *Korea and Her Neighbours* (1897); *The Yangtze Valley and Beyond* (1899). She was the first English person to travel extensively in the interior of Japan, or to cross the mountain ranges in the south and east of Persia; she approached Tibet both from India and from China. Like all serious traveler-explorers she took time to understand the lands and the people she journeyed among—her journeys were seldom less than six months long, and often more. For the most part she traveled by horse or by boat, with one or two servants only, though sometimes she accompanied a military survey party. She was a sharp and sympathetic observer of landscape and cultures; she recorded measurements of climactic conditions, of river flow, and route conditions; she noted with enthusiasm the flora and fauna of the region.

Though she had no conscious idea of traveling on behalf of the Empire, yet her books were nevertheless part of the great Imperial project, and proved of considerable utility to those responsible for Imperial policy, from the prime minister downwards, who valued the accuracy and acuteness of her accounts of the remote but strategically important peoples she had visited. Similarly she never thought of herself as a geographer, yet she was recognized time and again by the Royal Geographic Society for the unique information she collected so precisely and embodied so vividly. She was by no means a politically active feminist, yet it was perhaps this recognition in exclusively male-dominated spheres, as a proof of what women *could* do, that she valued most of all.

As a woman indeed there were barriers of mistrust to be broken, and early in her life Bishop felt the need to appeal to male experts to corroborate the accuracy of her accounts; but other fundamental aspects of her personality led the establishment to accept her. She was a committed Christian, convinced that Christianity alone of all the world religions retained integrity and truth; she became active in work on behalf of missionary organizations —and yet she recognized remarkably clearly all the moral dangers of missionary activity, so often arrogantly imposing not just a faith, but a destructively inappropriate European system of living on cultures that were older than hers. She was at one with the majority of her countrymen in being also a thorough believer in the superiority of Western civilization. She was anxious, for instance, to see China move into the twentieth-century world of competitive international trade and rapidly developing technology with the same skill as Japan, but where she differed from many was that she believed this was best done from within the existing culture, adopting and adapting Chinese customs and beliefs wherever possible.

Against Bishop as a representative figure of the explorer-traveler-writer during the period I would like to pose Blunt. In all the respects mentioned he took the opposite view. He was a sexually voracious male. He was an agnostic who, as a result of a number of important expeditions in the Middle East, was drawn deeply into Islam, and valued it above other systems of life. In all his writings he utterly opposed imperialism, or indeed any political, cultural, or religious interference by Western powers in the affairs of other races. He was a political activist, writing, as well as his accounts of his explorations, incessant letters to the newspapers, and attempting, sometimes with success, to exploit his extensive political acquaintance. He even went to prison for his involvement in the Irish attempts to rid themselves of English landlords.

Bishop, Blunt, Kingsley, and George Curzon were primarily explorers; all set out to experience in far distant areas what very few of their countrymen or women were likely to experience, and their reports have the value for the reader that travelers' tales have had throughout historical time—a vicarious experience of the strange and the exotic, more or less powerful and vivid as the powers of the writer are greater or lesser. As the tours of Cook and his competitors funnelled more and more English people into Europe, there was less and less unfamiliar territory available so close to home, but there was still apparent a gradation in travel writing about Europe. There were certain areas still very little known in England, and the experiences of a writer such as Edith Durham, who explored the Balkans in the first decade of this century (*Through the Lands of the Serb*, 1904; *The Burden of the Balkans*, 1905; *High Albania*, 1909), have a close affinity with those of Bishop or Kingsley. Though she did not write nearly so well as either, her intimate observation of the complex of peoples in the Balkans was of material use to the politicians of the day.

Since it seems to many journeying writers (and their readers) that the surmounting of barriers and the negotiation of physical dangers are essential parts of the proper travel narrative, some who remained in familiar European territory decided to set up their own difficulties: Robert Louis Stevenson, traversing the Cevennes with a recalcitrant donkey, getting lost in inhospitable territory; Hilaire Belloc, walking in a straight line (or as nearly straight as possible) from Northern France to Rome, perversely attempting an impossible Alpine climb in order to preserve his line. Both *Travels with a Donkey* (1879) and *The Path to Rome* (1902) are interesting books, because their writers were intelligent and imaginative; but the artifice of the project in the first place leaves a hollow impression at the end of the reading experience, as if the exploit has been one of pure self-gratification and the narration of it a kind of super-journalism.

The journeys of Stevenson and Belloc provide a kind of framework for their wide-ranging historical and cultural discussions; but there are certain other of the finest travel books that encompass so wide a range of material without the self-imposed difficulties. One might instance Samuel Butler's

16

Frontispiece to *Travels with a Donkey in the Cevennes*, Walter Crane,
Pentland Edition, vol. 1 (1906).

Alps and Sanctuaries (1882), which illuminates a small, relatively unpopular area of Switzerland and Italy, or Maurice Baring's accounts of Russia (*A Year in Russia*, 1907; *Russian Essays and Studies*, 1909), or George Gissing (*By the Ionian Sea*, 1901) and Norman Douglas (*Old Calabria*, 1915) in Calabria (the southernmost region of mainland Italy).

Arthur Symons, however, is a great travel writer of a different sort. He is most intensely moved not by the mountains and the moors, but by the artifice of towns. In the Dedication to *Cities* (1903), the volume in which Symons reveals most fully his gift for evoking the essence of urban experiences, he writes: "as love, or, it may be, hate, can alone reveal soul to soul, among human beings, so, it seems to me, the soul of a city will reveal itself only to those who love, or, perhaps, hate it, with a far-sighted emotion" (v). How this works in practice will be transparently clear to anyone who experiences the intensity of his accounts of Seville, a city he loves, or of Moscow, one he hates:

> To be in one of these hot and many-coloured rooms is like being shut into the heart of a great tulip. Only fantastic and barbarous thoughts could reign here; life lived here could but be unreal, as if all the cobwebs of one's brain had externalised themselves. . . . To live in Moscow is to undergo the most interesting, the most absorbing fatigue, without escape from the ceaseless energy of colour, the ceaseless appeal of novelty. (172–73).

Symons was a theorist and relatively minor practitioner of Symbolist verse; it seems to me that out of the *fin-de-siècle* synaesthesia that permeates such descriptions he is developing something more modern, that at times in these studies he creates what one might call post-Impressionist travel writing, in which the living object is seen most vividly in terms of form and color, distorted as in a Picasso painting for the sake of exposing underlying truths, and that in doing this he approaches greatness.

In the chapter on Rome we might compare two passages taken from his account of the Campagna.

> The beauty of the Campagna is a soft, gradual, changing beauty, whose extreme delicacy is made out of the action upon one another of savage and poisonous forces. . . . All the changes of the earth and of the world have passed over it, ruining it with elaborate cruelty; and they have only added subtlety to its natural beauty. (30–31)

This conjunction of beauty with poison, cruelty, and ruin is evidently decadent, Baudelairean, one might say—a throwback by 1903. But in

18

contrast, though retaining the same essential feeling, is the description of Lake Nemi:

> This space of dark water is closed in on three sides by tall, motionless cypresses, their solemn green, menacing enough in itself, reflected like great cubes of blackness, pointing downwards at the sky. The waters are always dark, even in full sunlight; they have always that weight upon them of the funereal trees which stand between them and the sun; and through the cypresses you can see Rome, far away, beyond the gardens, the stacked vines, the olive-trees, and the indefinite wilderness, set there like a heap of white stones. (34–35)

There is an Impressionist-Symbolist concern for atmosphere here, and a narrative force; but, allowing for the obvious differences between writing and painting, it is not just the phrase "great cubes," but the solid darkness of the water, the stacked vines, and Rome as a heap of stones, that point to the completion of a movement in English travel writing about the familiar places of Europe away from the detailed surface perceptions of John Ruskin toward an understanding and communication of essential form.

As further corroboration, here is a Venetian street-scene:

> [you] catch a glimpse of the heels of brilliant stockings, red, striped, white, occasionally a fine, ecclesiastical purple; now a whole flock of greenish yellow shawls passes, then, by itself, a bright green shawl, a grey, a blue, an amber; and scarcely two of all these coloured things are alike: the street flickers with colour, in the hot sunshine. (78)

Here are no people; they are simply indicated by passages of color in the form of stockings and shawls, in motion as much by the vibrancy of their contrasts next to each other, and the heat of the colors, as by the implied activity of their wearers.

Another significant element in *Cities* is the way Symons is perpetually flirting with the relation between East and West: in Seville it is the Moorish aspects of the city, in Venice it is St. Mark's; in Moscow it is the Imperial spoils, in Sofia it is the unfortunate mingling of two cultures. When he arrives, at the end of the volume, at the end of his journey from the heart of the Western world at Rome to the first outpost of the Eastern world in Constantinople, he confronts the question that the English, since their involvement in India in the eighteenth century, have frequently been forced to confront: what are we doing when we enter the Orient? It was a favorite question of explorers such as Richard Burton, Oswald Doughty, Blunt, Bishop, and Bell—one that became more immediate after England's assumption of *de facto* control over Egypt in 1882—but it was not so often asked (in public at least) by the indolent, rail-bound, urban aesthete. Oriental art is portable, and thus thoroughly at home domesticated in the aesthete's chambers; but it is stirring to find such a sensibility actually experiencing the East.

At the end of his first day, Symons writes, "I find myself bewildered, as if I have lost my way in my own brain" (214). At first and last his deepest feeling is that alienation experienced so powerfully by the Anglo-Indian administrators Kipling examines in his short stories:

> The sun soaks down into the narrow street, the smell of the mud rises up into your nostrils, mingled with those unknown smells which, in Constantinople, seem to ooze upwards out of the ground and steam outwards from every door and window, and pour out of every alley, and rise like a cloud out of the breath and sweat and foulness of the people. (216)

Symons's final summary is exemplary in its combination of attraction and alienation:

> The attraction of the East for the West is after all nostalgia; it is as if, when we are awakened by dreams, we remember that forgotten country out of which we came. We came out of the East, and we return to the East; all our civilization has been but an attempt at forgetting, and, in spite of that long attempt, we still remember. When we first approach it, the East seems nothing more than one great enigma, presented to us almost on the terrifying terms of the Sphinx. . . . Here everything is incommunicable; there is a barrier between us and them, as narrow perhaps and as real as the barrier between Europe and Asia: you only have to cross the Bosphorus. (259–60)

Cities is a vivid, lively example of what the characteristic aesthetic-Edwardian sensibility could produce when it turned to travel writing. Indeed it is perhaps not too much to say that the fulfillment of the almost insatiable demand for this kind of literature by so many significant writers led to the establishment of an idea frequently encountered in the imaginative literature of the period, coming to a climax in Wells: that a wider experience of the world will lead to greater understanding of the human condition, that it will generate a cosmopolitanism to counter the fiercely competitive nationalisms of the period (though some, like Forster in *Howards End* [1910], remained suspicious of the potential reductive effects of such cosmopolitanism).

Finally, it is important to mention the rapidly growing interest during this period in the emerging science of social anthropology, since writers such as Walter Spencer, Edward Tylor, and Sir James Frazer each drew upon and stimulated travel writers who were interested in reporting the social behavior of the peoples they encountered. Frazer's *The Golden Bough* (2 volumes, 1890; 3 volumes, 1900; 13 volumes, 1911–36) was the best-known work in social anthropology during the period. In fact, the title was ultimately somewhat misleading, for though the golden bough that Aeneas had to take to Persephone (and its connection with the mistletoe that alone could slay the Scandinavian hero Balder) was the starting point of his original investigation, by 1915, when the twelfth volume of his third edition was published, the book had become (as Frazer admitted) concerned with the more general question of the evolution of human thought "from savagery to civilization." It is hardly surprising that the work was widely read, since it is perhaps not too much to say that such a concern represented the single most important concept underlying the new Imperial project. *The Golden Bough* was also highly influential among imaginative artists. Angus Downie records (*Frazer and the Golden Bough*, 1970) that the first edition of the work "was read to Tennyson while his portrait was being painted by Watts. . . . Both poet and artist would seem to have been impressed, for the background of the portrait bears a pattern of mistletoe" (33). However, it took a while for the full impact of the work to be felt, and it is mostly in writing after the Great War that its effect is palpable; everyone knows how the book was a formative element on T.S. Eliot's *The Waste Land*, but Ezra Pound, William Butler Yeats, Edith Sitwell, Robert Graves, and D.H. Lawrence were all in one way or another directly influenced by Frazer. Of writers during this period, it seems probable that Joseph Conrad drew elements of his understanding of characters such as Kurtz (*Heart of Darkness*) and Jim (*Lord Jim*) from acquaintance with *The Golden Bough*.

3. The Literary Response

The difficult search during these forty-odd years would be for writers who did not in some way or another find ideas or the reality of abroad entering their work. Even taking into account the increase in the speed, variety, comfort, and availability of means of transport, it is surprising how many of the canonical writers from 1890 onwards seem to have been blessed (or afflicted) with a kind of travel-disease that kept them perpetually on the move.

In 1936 Ford wrote of himself in a note accompanying a portrait drawing by Georges Schreiber,

Thirteen times I have travelled the round that goes from London to New

York, New Orleans, the Azores, Gibraltar, Marseilles, Paris, London. . . .
If I had not so constantly travelled, I should have . . . written more and
better books; if I weren't, when travelling, constantly impeded by the
desire to settle down somewhere and start something growing and write
something, I should have travelled more happily and farther. (Schreiber,
Portraits and Self Portraits, 1936)

For some this invited a new cosmopolitanism; for others it reinforced
national feeling—though 1914 rendered cosmopolitanism much harder to
hold on to. Something similar might have been written by Rudyard Kipling,
Arnold Bennett, Hilaire Belloc, Maurice Baring, H.G. Wells, or Joseph
Conrad.

Imperialism and European politics and culture were, it is also true,
becoming year by year more urgently significant to a growing number of the
English. Thus, as I point out below (pp. 67–71), even the work of Thomas
Hardy, a novelist thought of as the essence of rural Englishness, is permeated with references to abroad, and when toward the end of the period
Forster writes his state-of-England novel *Howards End*, he includes implicit
and explicit contrasts with Germany and German imperialism.

Given the pervasiveness of the wider world in English literature of this
time, the hard task is to find adequate generalizations about the very many
different ways it is embodied. What follows is an account of some motives,
apparent or avowed, which seem to be those of a significant number of
writers. To begin with there is the fascination of what is different—a reason
for confronting foreign culture that has always stimulated writers. In this
period, however, in addition to the attraction of the exotic pure and
simple—Rupert Brooke imitating Paul Gauguin in Tahiti for instance, or
Frank Harris relishing bullfighting in Spain—there is more frequently than
in previous years an imperial dimension—Leonard Woolf administering
Ceylon or Kipling playing the Great Game in India. Kipling himself catches
precisely the feeling of the fascination of encountering whatever is Other in
his account of Japan:

> This good brown earth of ours has many pleasures to offer her children,
> but there be few in her gift comparable to the joy of touching a new
> country, a completely strange race, and manners contrary. Though libra-
> ries may have been written aforetime, each new beholder is to himself
> another Cortez. [2]

Much of the poetry that responds to the wider world during the period
belongs to this subdivision of the topic: aesthetes, decadents, Symbolists,
and Georgians were primarily concerned (to simplify greatly) with receiving

sensations and giving pleasure; hence it is the fleeting perception of a moment or of an emotion that Ernest Dowson, Arthur Symons, Rupert Brooke, and Lionel Johnson principally record in poems that bring abroad to their English audiences. Brooke's "Tiare Tahiti" (see p. 59 below) is a characteristic example, as is Dowson's "Breton Afternoon":

Out of the tumult of angry tongues, in a land alone, apart,
In a perfumed dream-land set betwixt the bounds of life and death,
Here will I lie while the clouds fly by and delve an hole where my
  heart
May sleep deep down with the gorse above and red, red earth beneath.[3]

Kipling is above all the poet who adds to this notation of difference a layer of specific concern, not just with experiencing but also of administering alien places and peoples. Any poem from *Departmental Ditties* (1890) or *Barrack-Room Ballads* (1892) would illustrate his early evocation of English life in India, but subsequently there was no area of the Empire that he did not write about with more or less insight.

The best of these poems of Imperial service, like "La Nuit Blanche" (1888) or "The Absent-Minded Beggar" (1899), go beyond simply communicating environmental or cultural difference; they make use of the difference to reveal fundamental insights into human nature. Conrad's *Nostromo* (1904) might be taken as an example of an extended work in which a sense of foreignness dominates the reading experience, but is ultimately recognized as being secondary to the main purpose. The subject of the novel is in the end not Costaguana itself or its diverse population, though Conrad spends much time and energy recreating the topography, society, and history of the fictional South American country. The subject is human nature's response to certain kinds of pressure, and a South American state has some attributes essential for Conrad's purpose that Victorian England does not have: political instability, economic and industrial backwardness, a colonized past but current independence, a substantial racial mix, and an exploitable resource of precious metal. There is considerable and important pleasure for the reader in experiencing the richness of Conrad's Costaguana, but in the end it is the development of the individual responses of the Goulds, Decoud, Mitchell, Monygham, the Avellanoses, the Violas, and Nostromo to the political, social, and topographical conditions of Costaguana that is important, and this development could have been charted in any region that could have been made to meet Conrad's requirements.

For another variation on this theme, one might briefly consider Ford's *The Good Soldier* (1915). It is not that much of the action has to take place abroad, but rather that Ford embodies in the novel one of the Victorian and Edwardian clichés concerning abroad—that it is proper to go abroad in order to conduct adulterous affairs. Edward Ashburnham runs the full house, the Indian affairs with bored expatriate wives that Kipling pins down, the expensive Riviera affairs with dancing girls, and the hothouse affairs in German spas. It is the climactic, the serious affair of Ashburnham's life (if we are to believe the narrator Dowell) that is the essentially English one.

In "The English Flag" (1891) Kipling wrote "what should they know of England who only England know?" He was concerned with the narrowness of vision of the "poor little street-bred people" who seemed to him to undervalue the Empire, but his phrase points to a reason why several other writers brought abroad into their work. For instance, George Gissing in *The Emancipated* (1890) (see below, pp. 76–79), or Forster in *A Room with a View* (1908), send characters to Italy because what they experience there (and could experience precisely in that way nowhere else) will change the way they view England and their lives at home.

A Room with a View embodies the conflict between the artifices of Edwardian society and the naturalness of Italian experience for the soul of Lucy Honeychurch, the young, commonplace, more-or-less unformed heroine of the novel. While in Italy she observes a murder and is kissed amidst a cataract of violets on a hillside in view of Florence; both are seminally formative violent events for her. The Italians involved in these events act with passion and love; the English, with the exception of George Emerson, the man who rescues her from the murder scene and embraces her among the flowers, react frigidly and with fear.

The second part of the novel takes place in Surrey, and we see Lucy trying, because of fear, not to understand the lessons Italy had been teaching her, pretending to herself that she could marry the conventional, objective Cecil Vyse, repressing her feelings for George. This attempt ends in her lying to everyone, including herself, in order to retain her home-counties, middle-class worldview. It is only thanks to the momentary defection of a chief exponent of this worldview that she is enabled to recognize the truth, and begin to live—as George's wife. The novel appropriately ends with their honeymoon in Florence.

The commitment of Italians to love and to people rather than to ideas and things is offered as a model for the English, who shelter themselves from experience and each other in ice-houses of customary behavior. In the terms proposed by *Howards End*, Lucy's specifically Italian experiences were instructing her how to connect.

Other writers make equally important use of a foreign perspective from which to see England more clearly, but are more interested in the effects of remoteness by themselves than in any contrast between England and the specific culture and environment of the foreign viewpoint. As an instance of this one might take Virginia Woolf's *The Voyage Out* (1915). The foreign town Woolf created, somewhere in South America, for *The Voyage Out* had to be within easy reach of the kind of wilderness not found in Europe, but otherwise it might have been put down on any more or less tropical coastal fringe—a sharp contrast in this respect to *Nostromo*. What it is that Woolf wishes to show us about England from this imprecise but remote point of view is suggested when her novel is read as a direct response to Conrad's *Heart of Darkness* (1902). Though Conrad's narrator Marlow is unequivocal in his disgust at the Belgian colonial exploitation in the center of Africa, he and the frame-narrator are at best ambiguous about English Imperialism. By echoing and even parodying *Heart of Darkness*, Woolf indicates that greed, cynicism, and sentimentality are at the root of the English colonial enterprises, as much as those of any other nation.

The first part of *The Voyage Out* is the voyage itself, from the Thames to anonymous South America, with a stop at Lisbon. Before Woolf's steamer, the *Euphrosyne*, leaves the Thames, there is offered a vision of London, culminating in this sentence: "From the deck of the ship the great city appeared a crouched and cowardly figure, a sedentary miser" (12). In *Heart of Darkness* there is a vision with a different emphasis: "The air was dark above Gravesend, and farther back still seemed condensed into mournful gloom, brooding motionless over the biggest, and the greatest, town on earth" (45).

The *Euphrosyne* stops en route at Lisbon, and a globe-trotting couple hitch a lift—Richard (who is a politician) and Clarissa Dalloway. Once at sea again Woolf gives a conversation between husband and wife that is full of ironic echoes of the first pages of Conrad's narrative:

> "Being on this ship seems to make it so much more vivid—what it really means to be English. One thinks of all we've done, and our navies, and the people in India and Africa, and how we've gone on century after century, sending out boys from little country villages—and men like you, Dick, and it makes one feel as if one couldn't bear *not* to be English! . . . ."
>
> "It's the continuity," said Richard sententiously. . . . He ran his mind along the line of conservative policy, which went steadily from Lord Salisbury to Alfred, and gradually enclosed, as though it were a lasso that opened and caught things, enormous chunks of the habitable globe.
>
> "It's taken a long time, but we've pretty nearly done it," he said; "it remains to consolidate." (53)

The *Euphrosyne* encounters two English warships, as the vessel on which Marlow travels to the Congo comes across a French warship. The tone of the narrator's description of the English battleships is distinctly at odds with the response of the Dalloways:

She had sighted two sinister grey vessels, low in the water, and bald as bone, one closely following the other with the look of eyeless beasts seeking their prey. . . .
Richard raised his hat. Convulsively Clarissa squeezed Rachel's hand. "Aren't you glad to be English!" she said. (75)

The final detail that clinches the connection between the first part of this novel and the first part of Conrad's is an account Woolf gives of an early attempt by the English to colonize the particular part of South America that her travelers have reached:

All seemed to favour the expansion of the British Empire, and had there been men like Richard Dalloway in the time of Charles the First, the map would undoubtedly be red where it is now an odious green. But it must be supposed that the political mind of that age lacked imagination, and, merely for want of a few thousand pounds and a few thousand men, the spark died that should have been a conflagration. (101)

The "spark" is quoted from Conrad's first narrator's description of Elizabethan adventurers bearing "a spark from the sacred fire" (47), while the sarcastic reference to the map echoes Marlow's probably unironic analysis of a similar map in the Brussels headquarters of the company that is to employ him (55).

But, as in *Heart of Darkness*, the voyage out is only the prelude to a more metaphorical/symbolic one. The river journey provides Marlow with a point of view sufficiently alienating to allow him to consider himself and Western civilization from scratch. Woolf's central character, Rachel Vinrace, has more in common with Forster's Lucy Honeychurch than with Marlow, and her voyage is that more frequently encountered one of crossing the shadow-line; and yet it also contains a journey by steamer up a tropical river. She travels with the man she is perhaps coming to love, and a group of others, up the local river to an Indian village. Once again Woolf turns to *Heart of Darkness* for an appropriate context: the man (Mr Flushing) who organizes the journey is an exploiter of the natives in a small way, purchasing clothes and ornaments; as his wife says: "My husband rides about and finds 'em; they don't know what they're worth, so we get 'em cheap. And we shall sell 'em to smart women in London" (286). After the voyage begins "they seemed to be driving into the heart of the night, for the trees closed in front of them, and they could hear all round them the rustling of leaves" (325). They encounter a clearing in the forest and a hut where a famous explorer died; Hirst says "It makes one awfully queer, don't you find? . . .

26

These trees get on one's nerves—it's all so crazy. God's undoubtedly mad. What sane person could have conceived a wilderness like this, and peopled it with apes and alligators? I should go mad if I lived here—raving mad" (336).

When Rachel Vinrace goes out into the forest amidst all these Marlovian echoes, instead of encountering unspeakable rites, she finds that she and Terence Hewet are in love. Later they imagine by contrast life in England: "there would be English meadows gleaming with water and set with stolid cows, and clouds dripping low and trailing across the green hills" (367). "Lord, how good it is to think of lanes, muddy lanes, with brambles and nettles . . . there's nothing to compare with that here—look at the stony red earth, and the bright blue sea, and the glaring white houses—how tired one gets of it!" (368). Woolf, it seems, sends these people out to South America so that the sentimentality of their evocation of rural and urban England will be by the environmental contrast more evident. But a little later Rachel alters the whole perspective:

> "What's so detestable in this country . . . is the blue—always blue sky and blue sea. It's like a curtain—all the things one wants are on the other side of that. I want to know what's going on behind it. I hate these divisions, don't you, Terence? . . . Just by going on a ship we cut ourselves off entirely from the rest of the world. I want to see England there—London there—all sorts of people—why shouldn't one? why should one be shut up all by oneself in a room?" (369–70)

Suddenly her desire is not for "great church towers and the curious houses clustered in the valleys," not for the conventional exile's view of home, but for something far more inclusive: for the power to transcend the limitations of time and space and the isolation of the individual, the supernatural power to be everywhere and in all people. When, soon afterwards, Rachel develops typhoid, the isolated enclosure in a room she had metaphorically invoked becomes literal.

Marlow claims to understand Kurtz partly because he follows him to that ultimate point of no return at which Kurtz is enabled to utter his famous judgment; we too follow Rachel through the pattern of her thoughts as she slowly dies, and we experience her death through Terence: "they seemed to be thinking together; he seemed to be Rachel as well as himself; and then he listened again; no, she had ceased to breathe. So much the better—this was death. It was nothing; it was to cease to breathe" (431). There is no horror, though perhaps there is release from the enclosing room. Until we die, we are all shut up by ourselves in a room, and whether it is in unspecified South America or on the too specific Thames Embankment, there is no escape from the room—certainly not into patriotic or sentimental generalizations about "home."

A number of poets have felt homesick like Terence, and it is to be suspected that Woolf would have treated Rupert Brooke yearning for Granchester from Berlin or James Elroy Flecker hating Grenoble as she treated Rachel, suggesting that their home thoughts would prove to be illusions. Nevertheless, the deep pain of betrayal in Flecker's "Brumana" (1916) does speak powerfully and eloquently for all those Englishmen who, seduced by prospects of "older seas, / That beat on vaster sands / . . . Lands / where blaze the unimaginable flowers," find themselves amid old and strange and exquisitely beautiful things and yet can only try "to forget the wandering and pain" and "dream and dream" that they are home again.

Woolf's novel was her first creative step away from Victorian realistic or naturalistic fiction, which was a reflection of the age's expanding mechanical materialism. Other writers found other ways of escaping from this dominant mode: the 1880s saw a growth in stature and significance of two literary approaches that oppose naturalism, the doctrines of art for art's sake (or aestheticism—one might think of it as European) and of romance (or adventure-fiction—which one might call Imperial or exotic). It was also during this decade that the new Imperialism began to take popular hold. All three developments stress idealism at the expense of realism or naturalism, or the idealism that must be seen as lying behind the realism (though once you have said so much, aestheticism begins to move rapidly apart from romance and Imperialism, with which it has for the while felt very uncomfortable).

Robert Louis Stevenson and Rider Haggard were the pioneers of the revival of adventure-romance, and of these two, Stevenson had more to say, in a random way, about the theory of romance, which he saw, in moral and philosophical terms, as a counter to scientific materialism. Adventure stories are quest stories, and the object of the quest is an ideal unfound in the daily life of the questers; the quest does not have to be abroad, but the journey is an important element, and the encountering of the unfamiliar. If the romance remains predominately in England (or Scotland), then, in order to produce for the reader that sense of distance, it almost *has* to be set in the past. Adventure-romance is a late-nineteenth-century restatement of Romanticism: it offers hope from the operation of chance, not despair; it offers individual freedom, not social binding; it offers the unpredictable, not the inevitable; and though even the achievement of the quest may not encompass the attainment of the ideal, or not for all the seekers, the perception of the ideal leaves its mark. Such freedom and unpredictability are inevitably more easily found abroad, and even when the reaction to late-Victorian social realism came in terms of an exploration of the power of personal relationships to transcend social constriction and scientific materialism,

28

"I saw the fire run up her form," Maurice Greiffenhagen, illustration
to Rider Haggard's *She* (1913).

writers still found it necessary to provide a foreign perspective for their characters (as in Forster's *A Room with a View* or Lawrence's *Women in Love*, 1921).

Though adventure-romance is essentially a fictional genre, Flecker's "The Ballad of Iskander" (*Collected Poems*, 1916) provides a model for the theory just outlined. In form it is squarely in the tradition of the English literary ballad, but Flecker exoticizes the poem by giving his characters Arabic versions of their names (the ballad's first two lines read: "*Aflatun and Aristu and King Iskander / Are Plato, Aristotle, Alexander*"). King Iskander, "King / Of Everywhere and Everything," orders the Lord of his ships to gather twenty men who know all tongues, impress Aflatun and Aristu, fit out a silver-mailed ship with a crew of fifty, and sail southwest in search of new peoples for the King to conquer. On the hundredth day out they are driven by a storm into unknown seas. They sail on for twenty-one years, their ship tarnishing and rotting, without meeting another vessel or seeing land; ultimately, though, they do sight a ship, which rapidly closes with them. This ship seems to the crew to look very much the way their own vessel did on first setting sail, and when questioned, the captain of the other ship says that "*SULTAN ISKANDER* sent us forth." Aristu the materialist suggests that the King must have grown impatient and ordered out a second ship in search of "Unconquered tracts of humankind," but Aflatun, laughing, calls him a fool, for Aflatun has recognized that they have encountered their ideal. He points out two "tall and shining forms" on the other ship, saying, "they are what we ought to be, / Yet we are they, and they are we." On these words a breeze stirs, and the silver and the blackened ships are driven in on one other, the black ship crumbles into air, and the silver ship is left to sail on for eternity.

The poem is an archetype of the adventure-romance quest, the achievement of which leads to consummation with the ideal; of course it is far beyond most human searches, which end even at their most ideal, like that in Haggard's *She* (1887), with the notation of human failure before the eternal; it is the journey, not the arrival, that is most important. Even scientific romances, like those of Wells or of Arthur Conan Doyle, conform to this model, since their given assumption is one of advance into the unknown beyond the established material laws, wherever there seem to be available loopholes.

With the nearer approach of universal education since 1870, a new readership grew up during this period, comprising what Whelpdale in Gissing's *New Grub Street* (1891) called "the quarter-educated; that is to say the great new generation that is being turned out by the Board schools" (496). This fresh addition of hundreds of thousands of readers to the market stimulated the production during this period of romances and adventure stories designed more or less specifically for them, or for children, by writers such as Hall Caine, Marie Corelli, Anthony Hope, Edgar Wallace, Maurice Hewlett, S.R. Crockett, G.A. Henty, Stanley Weyman, Marion Crawford (like Henry James, a Europeanized American). To generalize, these are often set outside England for the reasons already offered above; they are for the most part highly readable, sometimes intelligent, occasionally witty, usually totally Anglophile, and frequently racist; and they sold by the hundred thousand. The new millions of readers liked, as Gissing knew, to read not about themselves and their social and financial problems but about exotic places and daring deeds. They also liked (to adapt the narrator of *Heart of Darkness*) the meaning of the story, if there was one, to be like a satisfying kernel at the end of the crack-shell process of storytelling. This is why Conrad was not popular, and why Caine was.

Many of these popular romances were set in some remote English-administered corner of the world, and helped to shape contemporary attitudes to Empire. In his *Experiment in Autobiography* (1934) Wells, looking back over fifty years, tried to recall what it felt like to be an English schoolboy in the 1880s:

> It was made a matter of general congratulation about me that I was English . . . a blond and blue-eyed Nordic, quite the best make of human being known. . . . We English, by sheer native superiority, practically without trying, had possessed ourselves of an Empire on which the sun never set, and through the errors and infirmities of other races were being forced slowly but steadily—and quite modestly—towards world domin-ion. (99)

Wells was writing here from beyond the First World War and with rumors of the Second already in his ears, and so the palpable ironic tone is inevitable, and one can't be quite sure how much of this is necessary exaggeration after the fact. But still it may stand for a basic stratum of the English collective unconscious which, when brought to consciousness by any individual between 1880 and 1914, would provoke fierce responses of confirmation or disavowal.

Rudyard Kipling, Rider Haggard, Saki (Hector Hugh Munro), and John Buchan: the names of the imaginative advocates of the Empire as the key to England's moral and physical survival are well known; much less familiar are those who were hostile to it. Blunt loathed what England had done and was doing in the name of Empire to the Arab world—in particular to Egypt, where he had a home—and to Ireland. His diaries are a marvellous record of his rational hatred, but here it is appropriate to quote a few fragments from his poem *Satan Absolved* (1899). Blunt imagines Satan on a return visit to

31

heaven pointing out to God what an utter mess the world is in through the activities of man, activities that Satan had tried to warn God about often enough before. Particularly destructive are the works of the Englishman, who, Satan says ironically:

> goeth among the nations . . .
> to spread Thy truth, to preach Thy law of patience,
> To glorify Thy name! Not selfishly, forsooth,
> But for their own more good, to open them the truth,
> To teach them happiness, to civilise, to save,
> To smite down the oppressor and make free the slave.
> To bear the "White Man's Burden," which he yearns to take
> On his white Saxon back for his white conscience' sake.
> Huge impudent imposture! (*Poetical Works* 2.281)

Satan then points out that there were once places on Earth where the people lived in harmony with their environment like prelapsarian Adam and Eve; the first English attack on their happy state comes from "the white gospel-lers, / Our Saxon mission-men black-coated to the ears," who bribe and frighten the "natives" into Christianity. Having prepared the way, the missionary is followed by the trader, with his own gospel of free trade. Satan's account of the way he works is classic: he creates in them a desire for what they do not need—"cloth, firelocks, powder, rum"—and then, once they are hooked, he teaches them the meaning and consequences of western trading ethics. The imagined dialogue goes like this:

> "How shall we bring the price, since ye give naught for naught?
> We crave the fire-drink now." "Friends, let not that prevent.
> We lend on all your harvests, take our cent per cent."
> "Sirs, but the crop is gone." "There is your land in lots."
> "The land? It was our fathers'." "Curse ye for idle sots,
> A rascal lazing pack. Have ye no hands to work?
> Off to the mines and dig, and see it how ye shirk." (283)

Thus, Satan says, the white man enslaves and betrays, until in desperation the natives turn on the whites:

> Then loud the cry goeth forth, the white man's to each friend:
> "Help! Christians, to our help! These black fiends murder us."
> And the last scene is played in death's red charnel house.
> The Saxon anger flames. His ships in armament
> Bear slaughter on their wings. (284)

And, Satan says, the final bitterness is that the English conduct this slaughter in God's name. As a conclusion, Blunt can't resist a last stab at Kipling:

32

Their poets who write big of the "White Burden." Trash!
The White Man's Burden, Lord, is the burden of his cash. (284-85)

Conrad's attitude toward Imperialism in *Nostromo* and *Heart of Darkness* is explored in essays that follow; here it is worth mentioning *The Inheritors* (1901), one of his collaborative works with Ford, in which there has been put into effect a scheme for colonizing Greenland and bringing the Eskimos all the advantages of civilization: the Duc de Mersch, who has established the *Système Groënlandais*, anxious to get British funding for the Trans-Greenland Railway, talks about his colony thus:

> Progress, improvement, civilisation, a little less evil in the world—more light! It was our duty not to count the cost of humanising a lower race. Besides, the thing would pay like another Suez Canal. Its terminus and the British coaling station would be on the west coast of the island. (81)

Eventually the truth about the colony emerges: "flogged, butchered, miserable natives, the famines, the vices, diseases, and the crimes" (183); the similarity to Marlow's experience of the Congo is obvious.

Much of the writing about the Empire was based on the premise that violence was part of the imperial process. Warfare was usually found essential to prepare other races for English civilization, and it was often thought to be fun as well, so long as you didn't yourself get killed. In Europe, however, the balance-of-power ballet maintained thirty years of continuous peace, so that people began to think it was the natural state of affairs. It is thus somewhat surprising to find that these years saw the growth and flourishing of what is best called invasion literature. The prototype of these narratives was Sir George Chesney's *The Battle of Dorking*, which was published in 1871, the year the Prussians crushed France. This account of a successful German invasion of England was followed by fifty or sixty further imaginative descriptions of abroad meting out to the English at home what the English dealt in the colonies. This volume suggests something about the deep-seated anxieties of a country that had not been successfully invaded by a foreign power since 1066. Most writers saw Germany as the aggressor, but France and Russia had their supporters. However, the most exotic, the most famous, and, with Eskine Childers's *The Riddle of the Sands* (1903) (see below p. 45), the best written of the invasion stories was Wells's *The War of the Worlds* (1897), in which Mars was the enemy. The novel is also a potent indictment of imperialism, though it is not clear how many contemporary readers saw an analogy between the way the Martians sought to eradicate the English and the way the English had dealt with, for instance, the Maoris.

Wells went on to predict in various works (*Anticipations*, 1901; "The

33

Land-Ironclads," 1903; *The War in the Air*, 1908; *The World Set Free*, 1914) most of the significant tactical and technological developments used in the First and Second World Wars—developments that would make war increasingly destructive and less personal. However, when in 1914 the war itself arrived, his first response was to write the pamphlet the title of which provided a notorious catchphrase: *The War that Will End War*. Wells is only one example—Arnold Bennett, Ford Madox Ford, John Galsworthy, May Sinclair, and John Middleton Murry are others—of writers who were apparently irresistibly drawn into the creation of propaganda during the first year of the war, primarily as a response to the celebrated Wellington House meeting of September 1914, at which over twenty-five authors of note agreed to commit themselves to write for England's war effort. The Great War, for a while at least, brought most writers to view abroad through similar lenses. It was very hard for anyone writing about the war in a novel published before it had ended to be otherwise than patriotically enthusiastic about England's part in the conflict. An example that comes close to being evenhanded is Sinclair's *The Tree of Heaven* (1917). Sinclair sees the war as the culmination of social and political developments from Victorian and Edwardian self-satisfaction and the cultural turmoil of the years after 1910 (the year in which Virginia Woolf was later to claim the twentieth century began), suggested in the novel by Suffragism, Imagism, and Vorticism. When war is declared, only Michael, the poet in the Harrison family, resists the overwhelming pressure to enlist, and Sinclair's account of the reasons he and his fellow artists offer for resistance is convincing:

> They believed with the utmost fervour and sincerity that they defied Germany more effectually, because more spiritually, by going on and producing fine things with imperturbability than if they went out against the German Armies with bayonets and machine-guns. (*The Tree of Heaven* 330)

Though, with authorial approval, Michael ultimately repudiates his antagonism to the war as cowardice, the reader cannot so easily acquiesce in such a judgment. Almost everyone in the novel who goes to France dies; they almost all know that they will die, and it is evidence of Sinclair's honesty that she manages to convince almost all the time that her idealist view of the conflict is a viable one. She openly acknowledges the brutality and the vileness of the living and the dying, but she sets against the horror an almost religious exaltation of spirit which has value, and which it may not be possible to experience in any other way. For the most part the world nowadays is wary of claiming connections between spiritual exaltation and killing, but Sinclair is no propagandist in this novel. Her experience at the

34

front in Belgium lends authority to her claim that parents could hear of their children's deaths with pride, and that children could, even as late as 1916, be pleased to go to Europe to face death.

4. The Literary Response Region by Region

France Among the significant dancers in the European power ballet, France was unique in being a republic—a republic formed on the ashes of the humiliation of 1871. Margaret Oliphant's supernatural novel *A Beleaguered City* (1880) is an early example of imaginative work that examines some of the implications of republicanism. It is set in the provincial town of Semur-en-Auxois in the Côte-d'Or, soon after the fall of the second Empire, and the narrative raises questions about the place of religion in a democratic republic. However, it was neither republicanism nor the provinces that first came to mind when the English thought of France. For a while England and France were politically on bad terms, especially over imperial matters in the Far East, but this did not inhibit most middle-class Englishmen and women from recognizing Paris as the center of the artistic world and as the place above all that one should visit for a release from whatever Victorian conventions were galling; indeed in the 1890s and in the first decade of the twentieth century (the *belle époque*) the city may be said to have reached some kind of apogee as a pleasure palace. Symons began in 1890 to spend some time in Paris each year; he felt more at home in the artistic ferment and the music hall there (calling it the earthly paradise) than in Nineties London, though his Parisian poems seem hardly different from his London ones, embodying the same rhythms and the same images:

> My Paris is a land where twilight days
> Merge into violent nights of black and gold;
> Where, it may be, the flower of dawn is cold:
> Ah, but the gold nights, and the scented ways!
> ("Paris," *Poems: Volume One* 252)

It was easier for him, however (he implies) to be sexually promiscuous in Paris.

When Bennett in 1903 felt the need to escape further than London from his provincial background, he went to live in Paris (and ultimately married a French woman); several of his novels thereafter have scenes in Paris—*A Great Man* (1904), *Hugo* (1906), *Sacred and Profane Love* (1905)—but the finest is *The Old Wives' Tale* (1908), in which one of the wives in question is taken by her new husband on honeymoon to Paris in about 1869. She gets caught there during the siege in 1870, and surviving that experience, runs a hotel in Paris for many years. Bennett does two things in particular in this section of the novel. First he shows in detail what it means to go to Paris to escape England: one eats well, drinks champagne, gets Parisian clothes; Sophia from the Five Towns "put an arrogant look into her face, and thought of nothing but the intense throbbing joy of life, longing with painful ardour for more and more pleasure, then and for ever" (272–73). Bennett sets up a contrast between his heroine and a Parisian courtesan: Sophia is harsh, self-contained, provident; the courtesan, gradually descending to whore, is soft, weak, pitiable, self-indulgent. What clearer paradigm could one have for the English perception of the differences between themselves and the French?

By turning Sophia into the owner of a hotel in Paris, Bennett is able to glance also at the tourist from the point of view of the resident: "they all asked the same questions, made the same exclamations, went out on the same excursions, returned with the same judgements, and exhibited the same unimpaired assurance that foreigners were really very peculiar people. They never seemed to advance in knowledge. There was a constant stream of explorers from England who had to be set on their way to the Louvre or the Bon Marché" (394). In the end she sells the hotel and returns to Bursley, from which vantage point she reflects that by contrast with English businesses and shops the French were "so exquisitely arranged" and the people "so polite in their lying, so eager to spare your feelings and to reassure you. . . . She longed for Paris again. She longed to stretch her lungs in Paris" (449). Living and working in Paris had for a while lent provincial England the charm of remoteness, and it was surprising to her to feel that, after all, Paris was admirable, richer, freer, less egotistical than Bursley. Part of the difference she perceives is the result of a comparison of any provincial with any metropolitan society, but part is also the result of setting England alongside France.

Bennett (and Sophia) see Paris with vision blinkered by English provincialism, and their impression thus excludes any representation of Paris or Parisians on their own terms. It is very hard, in this context, to exclude from consideration Henry James's *The Ambassadors* (1903). Strether, the central consciousness of *The Ambassadors*, is decisively a New Englander, and there are some aspects of his response to Paris and his observation of the responses of his compatriots that make that wider stretch of water between origins and effects of the essence of the narrative. Nevertheless, James's lyrical and sensitive evocation of Paris as a place where townscapes and interiors give continual intense pleasure, and his depiction of people representative of Parisian culture and society, whose company can teach any outsider, English or American, a richer style of living, offer a liberal contrast to Bennett's harsh and limited materialism.

The endurance of Paris in the English imagination as the cosmopolitan world-center of artistic activity is testified to by Wyndham Lewis's *Tarr* (1918), the work of a revolutionary painter who was also a writer, containing an international cast of characters, set entirely in a Paris that is evoked in characteristically sparse sentences in the novel's first page:

> The Vitelotte Quarter is given up to Art: Letters and other things are around the corner. Its rent is half paid by America. Germany occupies a sensible apartment on the second floor. A hundred square yards at its centre is a convenient space, where the Boulevard du Paradis and the Boulevard Kreutzberg cross with their electric trams: in the middle is a pavement island, like a vestige of submerged masonry. Italian models festoon it in symmetrical human groups. (9)

For English men and women who merely sought foreign soil as an escape from too intimate domestic observation, Picardy or Normandy were but a short voyage away (in his 1904 essay "Montmartre and the Latin Quarter" Symons called Boulogne "the Englishman's part of France"). George Meredith (*Diana of the Crossways*, 1885) and Hardy (*The Hand of Ethelberta*, 1876) both allow heroines to make significant encounters with men on Norman beaches, but in *Beauchamp's Career* (1876) Meredith makes it clear in his account of the family of Renée de Croisnel that whatever English notions of the moral freedoms available in Paris, most of contemporary French life was in some ways more strictly bound by moral codes than the English.

The other great area of resort for the English in France was the Mediterranean coast, the Riviera, which for several generations the affluent English made in their own image. It is thus interesting that though Monte Carlo and gambling were topics in many novels, there was no rich or powerful imaginative evocation of the region; perhaps this was because so much of its raison d'être was money.

Spain Spain remained a European backwater, on the itinerary of only a handful of tourists; and yet, though Rome was the only Empire to which the English of this period habitually compared their own, there was also another model that they might have usefully considered, the rise and fall of the Spanish Empire. At the beginning of the period under consideration, the American W.H. Prescott published his *History of the Conquest of Mexico* (1878), and after 1900, R.B. Cunninghame Graham wrote a series of less academic accounts of fragments of the history of the Hispanic occupation of Central and South America, but still Graham was right to consider that Spain was neglected by English readers and writers. Of course no corner of Europe was exempt from the attention of the historical romancers, and Spain received its share of attention from them; but there are very few serious abiding works of fiction or poetry that draw inspiration primarily or partially from the country. Cunninghame Graham and Somerset Maugham wrote some brief stories or sketches, and Maugham wrote a travel book (*The Land of the Blessed Virgin*, 1905); Symons wrote a handful of poems about Spanish dancers and some of his best imaginative travel pieces were on Spain ("Montserrat" [1918], for example).

The one thing that most educated English people would have known about the Spanish was their passion for bullfights, and some of the best short fiction set in Spain concerns this occupation. Harris's story "Montes the Matador" (1891) presents the story of Montes, a dying torero whose whole success has depended on understanding the idiosyncracies of the bulls he is to face, as it is told to an English aficionado; though it is a life story, yet the narrator's fascination with the technique of the "sport" and his admiration for its practitioners shines through. Harris confessedly was attempting to improve upon Mérimée's *Carmen*, and consequently, perhaps for the first time in English fiction, there is in the story a whole range of archetypal Spanish characters.

To set against this celebration of skill, athleticism, and insight, there is Symons's piece: "A Bull Fight in Valencia" (1898, collected in *Cities and Sea-Coasts and Islands*, 1918), in which he is primarily concerned with *why* the bullfight is so riveting. He opens the piece thus:

> I have always held that cruelty has a deep root in human nature, and is not that exceptional thing which, for the most part, we are pleased to suppose it. I believe it has an unadmitted, abominable attraction for almost everyone; for many of us, under scrupulous disguises; more simply for others, and especially for people of certain races. . . . The problem is troubling me at the moment, for I was at a Spanish bull fight yesterday, the first I had ever seen; and I saw many things there of a nature to make one reflect a little on first principles. (131)

The late nineteenth century saw a tremendous growth in the academic study of primitive societies in what was to emerge as anthropology; in Symons's essay we have evidence of the growth of interest in imaginative literature in primitive emotions; this half-reluctant fascination with bullfighting points directly forward to Ernest Hemingway.

Italy Italy turned three faces to the writers of this period: one revealed the crumbling memorial of an Imperial past; another spoke of the abiding splendor of a past of the highest artistic greatness; the third offered the country and the people in the present.

38

As Imperialism became a vital force in English life, so Rome became more and more significant as a spoken or unspoken image in the collective consciousness of educated Englishmen (see for instance Stevenson in *The South Seas* [1896], quoted by Barry Menikoff below, p. 152). It was the most powerful previous empire with which they were familiar; and both its material decay and destruction, and the survival of part of its art and knowledge, provided material for thought. In particular the contrast was made, by those who were anxious for the further spread of English territorial domination, between the Roman simplicity of conquer, exploit, and rule with an iron fist, and the English complexity of infiltrate, annex, administer with justice, bring progress, enlighten with education, drag into the modern age, and if necessary bludgeon into submission in order to do these things. When, at the beginning of *Heart of Darkness*, Conrad's first narrator implied a distinction between the Roman occupation of England and the English occupation of its Empire, he was reflecting a widely held assumption:

What saves us is efficiency—the devotion to efficiency. But these chaps [Romans in England] were not much account, really. They were no colonists; their administration was merely a squeeze, and nothing more, I suspect. They were conquerors, and for that you want only brute force. . . . They grabbed what they could get for the sake of what was to be got. (50)

The second Italy, the Italy of the Renaissance and the Baroque, has always been at the heart of the English response to Italy, though there was a short break during the mid-Victorian period while Ruskin's medievalism held sway. Walter Pater's *Renaissance* was first published as early as 1868, and the work of J.A. Symonds reflected the continuing high regard in England for the art and literature of the period in Italy. George Gissing's *The Emancipated* (see p. 78 below) offers a good example of the way in which informed exposure to this astonishing exhibition can be transforming.

But it was by Italy in the present that most writers were moved. Samuel Butler, on the second page of his *Alps and Sanctuaries*, sounds the most common note: "But who does not turn to Italy when he has the chance of doing so? What indeed do we not owe to that most lovely and lovable country?" His rhetorical questions would have been echoed with personal modifications during this period by Gissing, Ouida (Marie Louise de la Ramée) James, Symons, Forster, Norman Douglas, Vernon Lee, or Lawrence. Italy was still the archetypal southern place: in his novel *South Wind* (1917) Douglas wrote of his imaginary Italian island Nepenthe:

"Tengia," Samuel Butler, from his *Alps and Sanctuaries* (1881).

40

Northern minds seem to become fluid here, impressionable, unstable, unbalanced—what you please. There is something in the brightness of this spot which decomposes their old particles and arranges them into fresh and unexpected patterns. That is what people mean when they say they "discover" themselves here. (200–201)

In the same way, Italians were the characteristic southern people. It was an idea familiarly found in Gissing, Symons, Forster, and Lawrence that southerners/Italians are open to life in a way that northerners/English are not. Their emotions were held to be less guarded, stronger, simpler. They seemed to be capable of greater joy and greater despair; they were more open and more various in their lovemaking (homosexuals such as Frederick Rolfe, Douglas, and Symonds found life more satisfying in Italy than elsewhere). Behind these perceptions lay the thought that Italians were more primitive than Victorians or Edwardians. Lawrence's suggestion in *Twilight in Italy* (1916) was that Italians were more in touch with themselves than the English—in his curious terms, that they were more closely attuned to the blood coursing through their bodies. The question is (as Kenneth Churchill in *Italy and English Literature 1764–1930* [1980], points out) what can the northerner actually learn from the southerner?

In *A Room with a View* (1908) Forster offers a lesson about the relative values of Italians and the English (see p. 24 above). He suggests that significant changes in the English are possible through exposure to southern society and environment, and Lawrence only takes this proposition a stage further, or I should say deeper; but as to becoming quasi-Italian, that is a different matter. Forster suggests in *Where Angels Fear to Tread* (1905) that it is not possible.

And there is another side to this lesson from the south. What writers also found when they visited Italy was that the state, having been unified and liberated, was attempting to catch up rapidly with France, Germany, and England; its industry was growing, its cities becoming modernized (compare Gissing's strong affection for Naples in 1886, expressed with slight undertones of concern at plans to put new boulevards through the heart of the old city [*lo sventramento*], with his sadness when he revisits it in 1899 when the avenues have begun to penetrate the city; and also with Symons's almost nightmare fear of the place in 1901). Ouida shows in *A Village Commune* (published in 1881 and unjustly neglected, despite her overenthusiastic prose and her propensity for black-and-white characters) that independence has permitted a new kind of corruption in local government which is even more destructive of the community and the well-being of the individual peasant than was Austrian, French, or Papal rule. These seem to the English visitor to be northern things invading the south, to be unmitigated horrors, de-

41

stroying what was fundamentally important for the Italophile, Gissing as much as Lawrence, appropriating Italy for their own empire-building purposes. It was shocking to find the natives anxious to follow the example the visitors were trying to escape from.

Though Italy was of importance to many of the writers during this period, for one or two it became almost an obsession. This is true of Rolfe, alias Baron Corvo, who spent the last five years of his life in Venice, and never left it, though if he had returned to England his extreme poverty would have ended. *The Desire and Pursuit of the Whole* (written in 1909–13, published posthumously in 1934) is a fictionalized account of those five years. It is an extraordinary book. Rolfe has a style, like Meredith or James, that is unmistakably and uniquely his own. It is permeated with neologisms, and frequently with the grammar and syntax of Italian rendered literally into English. There are strong elements of wish-fulfillment and of documentary reporting that leave the reader with passages to skip, but throughout there is an understanding of Venice and a love for the city that sustains the narrative at a very high level of intensity indeed. The city has two qualities in particular that attract Rolfe. The first is the characteristic Victorian assessment of Venice as a once-great city now in deep decay. The other is the sheer beauty of the whole environment. Rolfe is almost the only writer who seems to be at home with the Italians, to live with them not utterly as a *forestiere*, but communicating and understanding from within, able to pretend to be a tourist at need, but living and dying in this great good place.

Switzerland Getting to the remoter areas of Switzerland became much easier as the railways penetrated the mountains, and as alpine experience became more commonplace some of the romantic cult of the Alps was beginning to seem rather overdone. Butler was a deep lover of the Alps, yet by the time he was writing *The Way of All Flesh* (completed in 1885, but not published until after his death in 1903) he could not help but parody the conventional response of the romantic Englishman on first seeing Mont Blanc, or compare such a person's reaction to the Hospice at the Great St. Bernard ("The whole of this most extraordinary journey seemed like a dream, its conclusion especially, in gentlemanly society, with every comfort and accommodation amidst the rudest rocks and in the region of perpetual snow" and so on) with that of a contemporary, for whom the fame of the place has become by use less stirring to superlatives ("I went up to the Great St Bernard and saw the dogs" [46–47]).

Despite such ironies, the inbuilt post-romantic sublimity associated with mountains remained a potent force, and though high places are by no means confined to Switzerland, when writers during this period wished to tap this sublimity it was almost always to Switzerland that they turned—perhaps it

Date	Surname & Initials in BLOCK CAPITALS		Seat No
24/1/95	FINKELSTEIN, D		54

Please enter each title on a separate slip   Not more than four items will be supplied at one time

Newspaper Title	Place of Publication	Shelfmark (Overseas Titles)	Months & Years required
Times	London		Aug. 1877

Delivered	Not on shelf	Unfit for use	Not yet available

had something to do with the nation's long republican tradition. Meredith at the end of *The Egoist* (1879) places his enlightened lovers on a peak there, and so does Wells in *Ann Veronica* (1909). For such characters as Clara Middleton, Vernon Whitfield, Ann Veronica, and Capes, the free and pure altitudes of the Swiss Alps are the antithesis of the enclosed constraints of Patterne Park or Edwardian London. Their escape upwards offers them a different perspective on the world—quite different from, but analogous to those provided by the cultural escape to Paris or Italy, or the adventurous escape to Africa or the Antarctic.

Though still primarily a resort for walkers, climbers, and seekers after the sublime, Switzerland began also to have a reputation as a sanatorium, particularly for those with respiratory problems. Stevenson, for instance, wrote a few sketches about his experience of the mountain cure, which he found of little help, and Symonds managed to live on in Switzerland for a number of years.

Germany There was a transition during this period in commonplace English perceptions about Germany. The forging, under Prussian leadership, of a German Empire in Europe in the 1860s and 1870s, and the increasingly intense German ambitions for overseas possessions, gradually modified the previous pervasive sense of Germany as a land dominated by music, poetry, philosophy, and scholarship, a culture sustained by the patronage of a multiplicity of small princely courts. These developments also gradually destroyed the cousinly relationship between the two nations.

For an intense but fragmented English impression of German culture in the 1890s, one might turn to *Pointed Roofs* (1915), the first separately published, book-length section or "chapter" of *Pilgrimage* (published complete in 1938), a massive, essentially autobiographical novel by Dorothy Richardson. Her heroine, Miriam, goes from a middle-class family in financial trouble to teach in a school in Hanover that has both English and German pupils, something Richardson herself did in 1891. Richardson is a modernist writer, and her concern is with the development of Miriam, the novel-sequence's only consciousness; and thus, though vivid, her response to the country and its people is neither organized nor consistent. Here is one of Miriam's perceptions: "the Germans [were writing] with compressed lips and fine, careful evenly-moving pen-points; the English scrawling and scraping and dashing, their pens at all angles, and careless eager faces" (65). This might be thought a paradigmatic example, the Germans neatly constrained, precise, drilled, accurate, fine; the English random, energetic, outgoing, unreflecting. But on the other hand, when it comes to music, Miriam has already learned from watching and listening to the Germans playing, and from playing herself in Germany, that in music one can and should let oneself go, and if one does, then all will be different, real, magic music. So writing and music seem to cancel each other out, and ultimately to suggest that simple generalizations about racial characteristics are inadequate.

Elsewhere (148–49) Miriam compares German sentimentality with English irreverence; she feels that the German girls are much more fully content to be hausfraus than the English, despite the gross vulgarity of most German men. Above all she is excited by every difference, by being away from what is familiar: "the great Georgstrasse—a foreign paradise with its great bright cafés and the strange promising detail of its shops—tantalisingly half seen" (90). She is entranced by medieval architecture, and by her first experience of a Catholic church in the small town of Hoddenheim, and after following a heated argument between the English and German girls about whether England can show anything like it, she thinks to herself: "Of course there wasn't anything a bit like Germany in England. . . . So silly to make comparisons" (118). It is the difference that so moves her.

Miriam's experience offers nothing that a similar girl in a similar position might not have felt fifty years earlier; there is no awareness of German militarism or Imperial ambitions. However, some Englishmen began to be slightly nervous of Germany as early as 1871 when the Prussian armies so efficiently destroyed the French, though many supported their action against the second Empire. Over the years there grew an awareness, stimulated in part by writers like Meredith (see, for instance, *The Adventures of Harry Richmond* [1871] or *One of Our Conquerors* [1891], that Germans intended to be a dominant power not only in Europe but in the world, and that their economy was beginning to outperform England's. In *Howards End* (1910) Forster imagines a German (Schlegel) who fought for Prussia against Austria-Hungary and France, but who abandoned his country when he discovered that

> some quality had vanished for which not all Alsace-Lorraine could compensate him. Germany a commercial power, Germany a naval Power, Germany with colonies here and a Forward Policy there, and legitimate aspirations in the other place, might appeal to others, and be fitly served by them; for his own part, he abstained from the fruits of victory, and naturalized himself in England. (26)

Of course there is an irony reminiscent of frying pans and fires about such a move, an irony which Schlegel himself was aware of. He says to a German relation visiting London: "No . . . your Pan-Germanism is no more imaginative than is our Imperialism over here. It is the vice of the vulgar mind to be thrilled by bigness" (27). And yet the reader is forced to speculate that for the Schlegels there was some quality in the English, perhaps their long apprenticeship in Empire, which enabled them to be content in London.

Forster also comments in *Howards End*: "the remark, 'England and Germany are bound to fight,' renders war a little more likely each time that it is made, and is therefore made the more readily by the gutter press of either nation" (60). If such a war were to occur, then the general assumption was that England's defense against German aggression would be what it has always been, the girdling seas and the navy that policed them. However, it was not just inflammatory journalists who anticipated such a war; many thoughtful writers believed that a time would come when these ancient fortifications might not prove sufficient.

The narrator of Childers's *The Riddle of the Sands* makes the point most vividly through his description of the newly constructed Kiel Canal (opened in 1895):

Broad and straight, massively embanked, lit by electricity at night till it is lighter than many a great London street; traversed by great war vessels, rich merchantmen, and humble coasters alike, it is a symbol of the new and mighty force which, controlled by the genius of statesmen and engineers, is thrusting the [German] Empire irresistably forward to the goal of maritime greatness. (95)[4]

And a little later he gives his version of the threat this ambition poses:

our great trade rival of the present, our great naval rival of the future, she grows and strengthens and waits, an ever more formidable factor in the future of our delicate network of Empire, sensitive as gossamer to external shocks, and radiating from an island whose commerce is its life and which depends even for its daily ration of bread on the free passage of the seas. (97–98)

Gradually the narrator uncovers a German plan to launch an invasion fleet of shallow draught vessels that would lie hidden behind the Frisian Islands until the appropriate moment. The novel has continued in print in part because it is one of the classics of small-boat sailing literature, and in part because it is a good suspense story; but it is also a carefully thought out and by no means alarmist account of the danger posed to England by German expansionism.

Bennett and Belloc, too, in their journals and travel writing, both notice and hate what they see as Prussian militarism, the Catholic Belloc distinguishing between the peaceful (and predominately Catholic) Bavarians of the south and the coercive bellicose Prussians of the north. Ford, whose immediate ancestry was half-German, made an eloquent comment on Belloc's distinction in a single sentence from *The Good Soldier* in which the narrator, accompanying his wife who is taking the cure at Nauheim (Bad Homburg) and contemplating an excursion, comments that "[Marburg] has the disadvantage of being in Prussia; and it is always disagreeable to go into that country."

Scandinavia and the Polar Regions As is suggested below (pp. 83–95), the literature of medieval Scandinavia was an abiding interest for writers during the period, especially William Morris and Maurice Hewlett; and the plays of Henrik Ibsen and to a lesser degree of August Strindberg were deeply influential, not only in the revival of the English theater in the 1880s and 1890s, but across the spectrum of literature. The countries themselves (or their people) were, however, rarely embodied in imaginative writing save as places to go fishing in. The Arctic and the Antarctic were the last areas of the world open to European exploration, and for a while Captain Robert Falcon Scott's heroic failure to reach the South Pole before Roald Amundsen, a failure that might have been devised for a Georgian boys' adventure book, caught the popular imagination. It is with a similar awareness that the author is offering something extraordinary to the reader that Hardy, in both *The Return of the Native* (1878) and *Tess of the d'Urbervilles* (1891), enters the consciousness of birds who have migrated south from the Arctic to Wessex for the winter:

> After this season of congealed dampness came a spell of dry frost, when strange birds from behind the north pole began to arrive silently on the upland of Flintcomb-Ash; gaunt spectral creatures with tragical eyes— eyes which had witnessed scenes of cataclysmal horror in inaccessible polar regions, of a magnitude such as no human being had ever conceived, in curdling temperatures that no man could endure. (*Tess*, 397)

Eastern Europe The band of European territory from Poland in the north to the Balkans in the south was debatable territory for the English during much of this period. The Austro-Hungarian Empire was in decay, propped up with German support, and the Balkans were the the only area of Europe in which fighting actually took place before 1914. Nevertheless, they were remote from western off-shore islands, and the landscape and people were unfamiliar. The only way that Eastern Europe figured in imaginative writing was as a place of romance. Stevenson set his Meredithian *Prince Otto* (1885) on the Eastern edge of the German Empire, while Anthony Hope most successfully invented Ruritania in order to give Rudolf Rassendyl a chance to cover for *The Prisoner of Zenda* (1894) and to continue his fight with

Rupert of Hentzau (1898); all three feature a more or less generic landscape of pine forest and mountain, castles and hunting lodges, but have also railways and modern politics—a conventional mix of the medieval and the contemporary. The most celebrated of all Eastern European fictions during the period also combines these two time periods in a somewhat more sensational way. The investigation of vampires in a Transylvanian castle offers only a slightly exaggerated image of the English view of activities appropriate to this region, though it is fair to add that the author of *Dracula* (1897), Bram Stoker, was born in Ireland.

Russia In 1876, Russia was England's chief enemy, threatening Imperial interests all along its southern borders. By 1908 the perennially changing pattern of alliances meant that Russia was part of what was to become the Triple Entente, together with England and France. The country, nominally European and Christian, was nevertheless, for most readers in England, remote and unfamiliar enough to be chiefly a source of sensation and adventure, rather like the Empire. England itself was one of the homes of political emigrés from Russia, in particular of the Nihilists, and stirring political events in Russia's distant and recent past were the stock-in-trade of several romancers who, for a period of time, put out one or two titles a year such as *The Degenerate, A Village Temptress,* and *The Vortex.* (These three were all published in 1909 and written by Fred Wishaw, who grew up in Russia and began his career writing Russian pulp-fiction [mainly for boys] in 1895 with *Boris the Bear-hunter: A Tale of Peter the Great and His Times.* By 1915 he had published thirty-nine novels with Russian themes, thus demonstrating the demand for his product.) Henry Seton Merriman's *The Sowers* (1896) went through twelve impressions in its first year.

Russia was fast becoming more familiar to the cultural avant-garde in England through translations of first Ivan Turgenev, then Leo Tolstoy and Fyodor Dostoevsky, and also Nikolai Gogol, Alexander Pushkin, and Mikhail Lermontov. Writers in England of greater talent than Wishaw or Merriman were also drawn to the question of political action in Russia: Oscar Wilde wrote a play with a Nihilist heroine (*Vera; or the Nihilists,* 1880); William Morris and John Morley were members of the Society of Friends of Russian Freedom; Harris wrote a long story, "Sonia" (collected in *Montes the Matador,* 1900), about the anarchist attack on Alexander II, in which there is a continual contrast between safe, cynical English materialism and dangerous, mercurial Russian idealism. Conrad makes a similar contrast between materialism and idealism the primary vehicle of the moral issues in *Under Western Eyes* (1911), about which he wrote in his 1920 Author's Note that it was "as a whole an attempt to render not so much the political state as the psychology of Russia itself" (7), and later he refers to his "general knowledge of the condition of Russia and of the moral and emotional reactions of the Russian temperament to the pressure of tyrannical lawlessness, which, in general human terms, could be reduced to the formula of senseless desperation provoked by senseless tyranny"(8). Others were less concerned with revolution: though he had never visited the country, Gissing sends the hero of *The Crown of Life* (1899) to Odessa to work for a while in great content. But the writer who was the most fully committed to Russia and who did most to help English readers to see the richness and variety of the Russian culture and landscape was Maurice Baring (*Russian Essays and Stories*, 1908; *The Russian People*, 1911; *The Grey Stocking and Other Plays*, 1912; *Fifty Sonnets*, 1915).

The Ottoman Empire and the Middle East Conventionally the "sick man of Europe," the Ottoman Empire was viewed with predatory eyes by all of the Great Powers. By 1919 its European provinces in the Balkans were independent, its North African provinces were under the rule of other empires, and its Arab provinces were split between England and France, leaving Turkey isolated within its national borders. Of the provinces it was the Arabic regions, together with the Nejd (most of modern Saudi Arabia) that most haunted the English imagination. Oswald Doughty (*Travels in Arabia Deserta*, 1888), Wilfrid and Anna Blunt (*Bedouin Tribes of the Euphrates*, 1879; *A Pilgrimage to Nejd*, 1881; *The Future of Islam*, 1882), and Gertrude Bell (*The Desert and the Sown*, 1907) all wrote movingly about their Arab experiences: the harshness of desert living, which brought life down to essentials, the predominating certainties of the Muslim religion, and the rootedness of Islamic culture all made a direct appeal to spirits in revolt against the rapidity of change, the growth of luxury, and the breakdown of religious faith in England. Jerusalem was no longer a goal for the adventurous literary traveller—Cook's ran tours there with great efficiency. Egypt, since 1882 under a *de facto* British protectorate, was similarly opened up to regular tourists, and its dependency, the Sudan, offered good territory for Imperial romancers, as Gordon lost and Kitchener regained control of the region. A.E.W. Mason's celebrated novel *The Four Feathers* (1901) is a representative example.

Africa It would be possible to distinguish between fiction that reflects English experience in East Africa, Southern Africa, and West Africa during this period, but the tenacious survival of the attitude that the whole of sub-Saharan Africa was the ultimate exotic environment, full of wild vegetation, wilder animals, dotted about with still wilder humans (or "semi-humans"), and thus really only suitable for narratives of adventure and romance, makes it reasonable to consider the response to these areas together.

Haggard is frequently thought of as the definitive writer in this respect, and in some ways he does conform to the stereotypes; but within the romance framework of his stories there is almost no aspect of the social structure of the life of South Africa that he did not touch, though his characters only find themselves in towns in order to leave as soon as possible. He writes of missionaries in *The Lady of Heaven* (1908), of Africans in *Nada the Lily* (1892), of Boers in *Marie* (1912), of English settlers in *Allan's Wife* (1893), and of English hunters *passim*, and though his primary concern in all these works is with a narrative of action, yet they do proceed from, and help to keep alive, the passionate remembrance of his relatively short experience as an administrator and farmer in South Africa. He finds it difficult to take a black person or nation on equal terms with white, and he finds it hard not to conventionalize female central characters as Lilith-like seducer-temptresses—indeed, as in *Nada the Lily*, he finds it hard to make even a Zulu central female character genuinely black. These are failings common to very many of his class and sex, but what makes Haggard interesting beyond his value as a representative figure are those moments when the adventure-narrative crystallizes into sometimes surprising insights into relations between the colonizer and the colonized (see p. 117 below).

Haggard was writing his African romances from the 1880s up to the First World War and beyond. John Buchan, who spent just about as long as Haggard as an administrator in South Africa, though twenty-five years later, was much more of an idealist. He wrote only one novel directly based on his experience of the country, *Prester John* (1910), though he also wrote a lyrically passionate account of the Eastern Transvaal in *The African Colony* (1903). *Prester John*, written for children, survives as a period piece. Like Haggard's romances it is an adventure story, but unlike them it has as central character an African, Laputa, who has had a Western education, and whose plot to sweep the white colonists from Southern Africa is only just foiled, thanks to the resource and courage of the young white protagonist. The white minority's half-fascinated fear of the black majority, not just in military but in social and cultural terms, which is a thread uniting very much fiction set in non-settlement colonies, runs strongly beneath Buchan's narrative. Buchan is as much interested in Laputa as he is in Davie Crawfurd, the young white hero of the book, and though the conflict is essentially the usual Imperial one between white and black, Buchan cannot help but show the human greatness of "Prester John." There is also a new development at the end of the novel, the product of Buchan's idealistic commitment to "civilization." Once the rising has been squashed, there is established a great native college at Blaauwildebeestefontein. Education, not violence, is the way forward for the African, he paternalistically implies.

It was almost impossible for writers in England during this period to

49

effectively represent the existence of African culture, or to imagine what it was like to be a black African—either a tribal African or a Western-educated one. Perhaps the best examples of the attempt, and the extent of failure, are Haggard's Zulus: they are full of white man's versions of Zulu thought, but they do notate a distinctive and coherent culture (if not civilization).

More typical is the representation of Africans in Edgar Wallace's "Sanders of the River" stories (1909); these narratives are strongly derivative from Kipling's Indian stories, stressing as they do the perpetual conflict throughout the Empire between the practical men who have devoted a lifetime to attempting to "understand the native" and the fresh-idealed theoreticians straight from "home" who attempt from ignorance to impose unworkable ideas. Wallace, however, unlike Kipling, cannot see any native as an individual; all are savage—though occasionally cunning—children and need to be treated as Victorians and Edwardians habitually treated children.

There is often a tendency to provide educated Africans with a biblical variety of English, but that is not unreasonable since most Africans who had learned English had learned it at missionary schools. An example is James Craven, an African who has a degree in medicine from Cambridge University, a central character in two of Mary Gaunt's romances set in West Africa, *The Arm of the Leopard* (1904) and *The Silent Ones* (1909). The first of these is directly analogous to *Prester John*, since Craven reverts to "savagery" and foments a rising against the English administrators, under pressure not of a social and political vision, but rather of sexual jealousy. At the last gasp Craven recovers his honorary white status by saving the man the attack was meant to eradicate; Buchan's Laputa, on the other hand, dies committed to his great dream. In the second novel Gaunt shows she recognizes the problems of alienation suffered by an African who has received the full blast of English education and culture, and is thus at home nowhere; but, like Haggard, her main purpose is the writing of exciting adventure, and she either cannot or will not take her isolated insight any further.

Conrad only wrote two very short works about Africa, and yet *Heart of Darkness* is in modern criticism the touchstone by which all other writing about Africa is judged. In this context it is notable that by choosing Marlow as his main narrator, Conrad excludes the African consciousness; it is debatable whether he did so because he wasn't interested in it, or because he knew he couldn't convincingly imagine or recreate it. There is much to be argued on either side of this question, but, without presenting the steps in the discussion, it might well be suggested that one of Conrad's aims in the novel was to procure, through exposure of Western civilization to the alien African experience, the reduction of all those who appear in it, of whatever race, to a bare undifferentiated humanity (see p. 123 below). It was a lesson that most English writers about Africa were not interested in learning.

India For the English, Kipling's is the voice of India during this period. The poetry, fiction, and journalism he published between 1886 (*Departmental Ditties*) and 1901 (*Kim*) brought home to the English audience with an unparalleled imaginative insight the nature of English service in India, both military and civil. In so doing he also gave them, through the acuteness of his observation and the intensity of his imagery, powerful sensory impressions of the variousness of the Empire and its people.

Some of his poems and stories are voiced for the ordinary soldier in the Indian army, and they are some of the most successful: the poems "Danny Deever" (1890), "The Widow at Windsor" (1892), or "Mandalay" (1890) on the one hand, and the stories "On Greenhow Hill" (1890) or "The Courting of Dinah Shadd" (1890) on the other. The remainder (save those in the two *Jungle Books*) are mostly narrated by an observer of the Anglo-Indian scene—perhaps a journalist—at home in the barracks, the officers' mess, and the bungalows and offices of the administrators of the Empire. The most enduring of these are stories that place the English into close relationship with Indians and note with tightly restrained clarity the consequences: "Without Benefit of Clergy" (1890), "Beyond the Pale" (1888), "To Be Filed for Reference" (1888).

Kipling shares with Hardy a view of the world that requires for the most part the detachment of the narrator of his fictions from the narrative itself, so that there is, in all the sharpness and intensity of the stories, an element of disengagement that underscores the pervasive sense that the English in India are working and living in an alienating environment. But as Hardy in part forgot his scepticism when writing *Tess of the d'Urbervilles*, so Kipling set his aside while writing *Kim*.

For once, in *Kim*, English rule is subordinated to the variety of India; it is a significant element in the multifariousness, but not the only one of significance, and India seems great enough to absorb it as it has absorbed other conquering and ruling races. Kim is at least half Irish (his mother's name and ancestry is not revealed to us) and totally British, and thus white and a *sahib*. This is an ultimately controlling perspective, but Kipling goes as far as he well can in overlaying this whiteness of birth with an altogether Indian nurture, so that Kim at first experiences the world from the underside of the class/caste system in Lahore, and later has to *learn* life from the English point of view. The contrast is made very vivid when the drummer-boy who has been assigned to keep an eye on Kim says "I'd run away if I knew where to go to, but, as the men say, in this bloomin' Injia you're only a prisoner at large"(145). Kim is free of all India, and it is part of Kipling's purpose to give his English or American audience some sense of what that might mean. The people closest to Kim represent almost archetypally as wide a range as possible of Indian racial and religious variation, and though

51

"Kim and His Lama," bas relief by Lockwood Kipling,
from Rudyard Kipling's *Kim* (1901).

he only travels in the northwest of the country, others, in particular the Lama who adopts and is adopted by Kim, travel to the east and to the south. Above all, Kim has the capacity for taking pleasure in each new experience, and that pleasure is infectious. Kipling's characteristic mode in *Kim* is the list; there are dozens of them through the novel:

> But it was all pure delight—the wandering road, climbing, dipping, and sweeping about the growing spurs; the flush of the morning laid along the distant snows, the branched cacti, tier upon tier on the stony hillsides; the voices of a thousand water-channels; the chatter of the monkeys; the solemn deodars, climbing one after another with down-drooped branches; the vista of the Plains rolled out far beneath them; the incessant twanging of the tonga-horns and the wild rush of the led horses when a tonga swung round a curve; the halts for prayers (Mahbub was very religious in dry-washings and bellowings when time did not press); the evening conferences by the halting-places, when camels and bullocks chewed solemnly together and the stolid drivers told the news of the Road—all these things lifted Kim's heart to song within him. (145–46)

There is nothing individually remarkable in such a list, but the cumulative rhythmical power of the details is almost overwhelming; and it is followed three short paragraphs later with another equally rich list in description of the lower Simla bazaar.

At no time before or since Kipling did an English writer (albeit one born, though not schooled, in India) force the metropolitan readership to consider so thoroughly what was involved in ruling an Empire that consisted primarily of geographically, climatically, and culturally remote areas. Forster's *A Passage to India* (published in 1924 and so properly outside the area of discussion of this volume) is a brilliant and deeply stimulating novel, but it did not command Kipling's immense audience, nor was it part of a fifteen-year sequence of writings which returned that audience repeatedly to the same issues. Furthermore, both the social and political developments in India during the gap of twenty years between *Kim* and *A Passage to India*, and the radical difference in outlook between Forster and Kipling, are embodied in the difference between quest and rape as underlying representational motifs.

Further south, Leonard Woolf's excellent *The Village in the Jungle* (1913) offers a convincing and moving account of peasant life in Ceylon. It is powerful evidence of Woolf's independence of spirit and openness of mind that his seven years' experience as a colonial administrator on the island did not prevent him from understanding and sympathetically interpreting the inhabitants, in a way administrator-writers about Africa could not.

The Lower Bazaar at Simla (1901).

54

China and the Far East Politically and economically, China was in process of being carved up by the European powers. It had proved under the Manchu dynasty to be incapable of assimilating Western influences and was thus vulnerable to military intervention. It was, however, missionary and, above all, trading activity that played the most important role in bringing China to English attention. In Meredith's *Beauchamp's Career* (1876), the conservative politician Tuckham dismisses China's attempts to exclude foreigners thus: "As to the case against the English merchants, the Chinaman is for shutting up his millions of acres of productive land, and the action of commerce is merely a declaration of a universal public right, to which all States must submit" (541–42). He provides in passing a piercing insight into the habit of mind ingrained by an empire dependent for its existence upon world trade, in contrast to one like the German Empire, which was based on an expanding continental territory. There is also an illuminating contrast between Tuckham's justification for economic penetration in China with that of an unnamed new-Imperialist trader in Gissing's *The Crown of Life* (1899): "I say that it's our duty to force our trade upon China. It's for China's good—can you deny that? A huge country packed with wretched barbarians! Our trade civilises them—can you deny it? It's our duty, as the leading Power of the world!" (254–55).

Imaginatively speaking, this is another region dominated for English readers by Conrad. *Almayer's Folly* (1895), "An Outcast of the Islands" (1896), *Lord Jim* (1900), "Falk" (1903), *Victory* (1915), *The Shadow-Line* (1916), and *The Rescue* (1920) are all set in what we now call Southeast Asia, though much of several of these stories is enacted at sea, as is almost all of a tale like *Typhoon* (1902). Conrad's East is essentially one of ports and coastline trading posts (especially Singapore, Bangkok, and on the island of Borneo), and with a few exceptions these places are populated by white sailors, traders, and colonial administrators, and by a mass of undifferentiated orientals. The coolies in *Typhoon*, "whirling like dust" after their possessions in the hold of the *Nan-Shan* during the hurricane, might be thought of as representative examples. Conrad shows how they and the sailors who come to sort them out (much as the European powers thought of themselves as sorting out China and its possessions) are mutually incomprehensible, but by brute force the sailors restore human order amidst the chaos caused by the elements. All the possessions of the Chinamen, including a quantity of silver dollars, are thrown out of the hold, and the Imperial metaphor of protection and exploitation seems complete—until at the end of the story the Ulsterman MacWhirr, the captain of the *Nan-Shan*, decides that the only fair thing to do is to share out the silver dollars evenly between the coolies, and we recognize that Conrad's metaphor is in fact one of the impartial justice of English Imperialism.

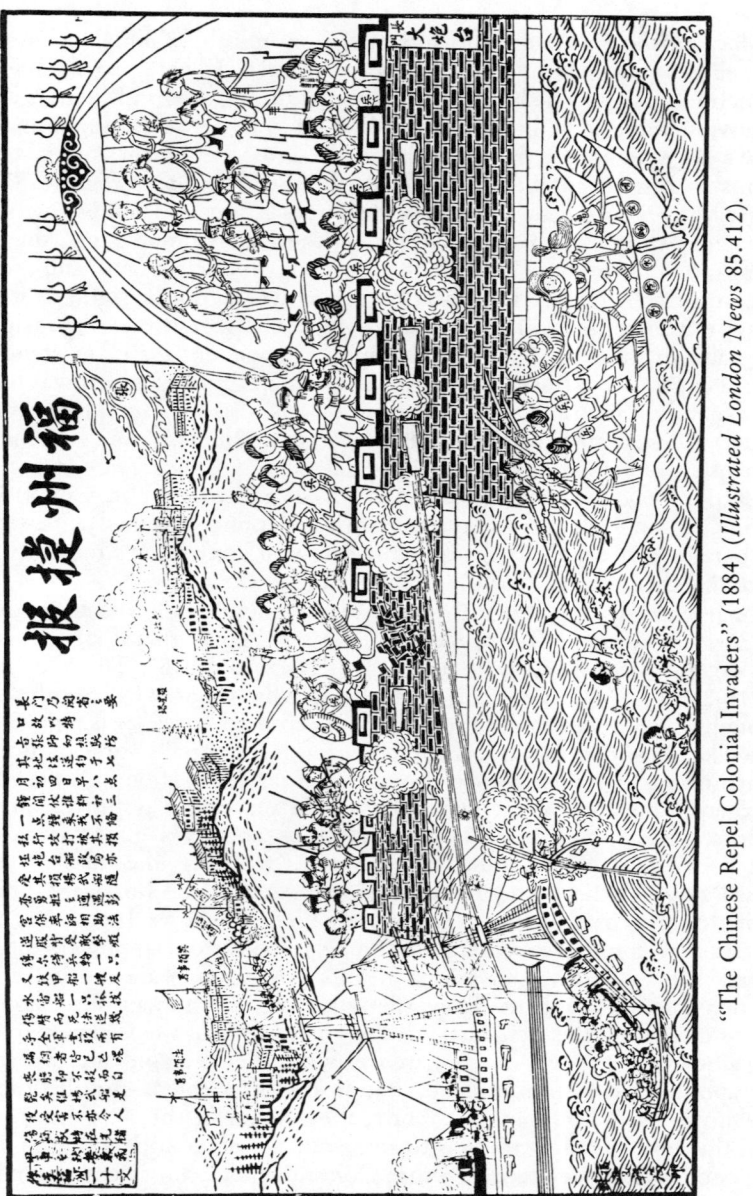

"The Chinese Repel Colonial Invaders" (1884) (*Illustrated London News* 85.412).

56

Bennett, in *The Old Wives' Tale*, shares Conrad's sense of the remoteness of Chinese culture from Western civilization; he says that the Frenchman Chirac "thought of women as the Occidental thinks of the Chinese, as a race apart, mysterious but capable of being infallibly comprehended by the application of a few leading principles of psychology" (377). A truer understanding of the Chinese people, as well as the Koreans and the Malaysians, was provided in the excellent travel writing of Isabella Bishop (see above p. 15).

Japan Japan at the beginning of this period was still relatively unknown; by 1906 it had become a favorite topic of speculation in Britain. The Japanese had just defeated one of England's most abiding enemies, the Russians, and this success, together with their position as a nation inhabiting a series of relatively small islands off the shore of a major continent, naturally gave rise to comparisons between the English and the Japanese, a comparison that was made more resonant by the rapidity with which the Japanese adopted and adapted economic, political, military, and occasionally cultural and intellectual theories from Europe and the United States. Even before this military success, Kipling wrote in 1889:

> Verily Japan is a great people. Her masons play with stone, her carpenters with wood, her smiths with iron, and her artists with life, death, and all the eye can take in. Mercifully she has been denied the last touch of firmness in her character which would enable her to play with the whole round world. We possess that—we, the nation of the glass flower-shade, the pink worsted mat, the red and green china puppy dog, and the poisonous Brussels carpet. It is our compensation . . .[5]

This is a fine insight; had he experienced as fully the "firmness" as he did the art, he might well have forecast accurately the future direction of Japan.

Kipling focussed on what was at the time the Western world's chief idea about Japan, excitement about the artifacts of their culture. Japan was one of the artistically fashionable foreign places during the period; not because many good writers or painters went there, but because first the art and then the poetry and drama of Japan had an important influence on a number of significant creators. Wilde and other British Nineties Symbolists brought Japanese details into their work partly through exposure to the paintings and decorative designs of James Whistler. Later, Yeats and Pound, through the work of Ernest Fennelosa, were able to study, value, and assimilate elements from Japanese Haiku and Nōh plays. But Japan as a place and a culture was a part of hardly any considerable literary work.

The Pacific and Australasia I cannot resist quoting, as a characteristic example of Edwardian Imperialist writing, the second paragraph of a travel book entitled *The Savage South Seas* (1907):

It is strange that so little is known of [the South Sea Islands], and that so few people have bothered themselves to visit them. A few missionaries, explorers, and adventurers have written about and spent a few months on them, but what is this when there are miles and miles of the most beautiful country crying out for people; there is wealth, both mineral and vegetable, waiting for the industry and enterprise of good men to reap, and above all, there is a delightful climate and a race of savages who in themselves repay the inconveniences of the journey.[6]

One writer who visited some of these islands and stayed, not for a few months, but for the remainder of his short life, was Stevenson. Out of this experience came two novels written in collaboration with Lloyd Osbourne, *The Wrecker* (1892) and *The Ebb-Tide* (1894), and a good novella: *The Beach of Falesá* (1892). Though the accounts of adventure on Midway Island in *The Wrecker* and on Tahiti and Attwater's island in *The Ebb-Tide* provide examples of the remotely exotic seen through pessimistic late-Victorian eyes, *The Beach of Falesá* may be taken as representative.

The novella has things in common with other stories that consider the life of an Englishman working in an alien tropical or subtropical environment; there are occasional reminiscences of Kipling and glances forward to Conrad. Stevenson's narrator, Wiltshire, is a half-educated trader, whose attempt to establish himself at Falesá provides the thread on which Stevenson hangs his examination of the practice and morality of such trade. Soon after arriving, Wiltshire goes through a fake marriage ceremony with a native girl, Uma, and comments:

A man might easily feel cheap for less. But it was the practice in these parts, and (as I told myself) not the least the fault of us white men, but of the missionaries. If they had let the natives be, I had never needed this deception, but taken all the wives I wished, and left them when I pleased, with a clear conscience. (254)[7]

Stevenson also allows him to condemn out of his own mouth the white man's attitude to the relationship of the native Kanakas and the law:

They haven't any real government or any real law, that's what you've got to knock into their heads; and even if they had, it would be a good joke if it was to apply to a white man. It would be a strange thing if we came all this way and couldn't do what we pleased. (272)

58

There is some exotic landscape and flora and fauna, and there is a strong sense of the beach-and-jungle-centered life, but the final impression is one of white supremacist futility. Wiltshire stands by his wife, Uma, and the story ends with this piece of introspection, which charts clearly the uneasy position a white man gets into when he allows common human responses of love and family feeling to interfere in his relations with the natives, but cannot at the same time shed his sense of inherent superiority.

I'm stuck here, I fancy. I don't like to leave the kids, you see: and—there's no use talking—they're better here than what they would be in a white man's country. . . . But what bothers me is the girls. They're only half-castes, of course; I know that as well as you do, and there's nobody thinks less of half-castes than I do; but they're mine, and about all I've got. I can't reconcile my mind to their taking up with Kanakas, and I'd like to know where I'm to find the whites? (336)

The fully Gauginesque exoticism of the Pacific islands surfaces in a few of the poems that Brooke wrote on a journey from Hawaii to New Zealand in 1913 and 1914. The finest of these South Sea poems is "Tiare Tahiti." It is structured around the proposition that though the wise say that after death we will be subsumed into the ideal ("And my laughter, and my pain, / Shall home to the Eternal Brain"), yet in face of the beauty and love and pleasure of life on this island "There's little comfort in the wise"; "there's an end, I think, of kissing, / When our mouths are one with Mouth" (with its deliberate allusion to Marvell's "To His Coy Mistress"). Tahiti is a Gauguinesque earthly paradise, and Mamua is Eve:

> *Taü here*, Mamua,
> Crown the hair, and come away!
> Hear the calling of the moon,
> And the whispering scents that stray
> About the idle warm lagoon.
> Hasten, hand in human hand,
> Down the dark, the flowered way,
> Along the whiteness of the sand,
> And in the water's soft caress,
> Wash the mind of foolishness,
> Mamua, until the day.
> Spend the glittering moonlight there
> Pursuing down the soundless deep
> Limbs that gleam and shadowy hair,
> Or floating lazy, half-asleep.
> Dive and double and follow after,

59

Snare in flowers, and kiss, and call
With lips that fade, and human laughter
And faces individual,
Well this side of Paradise!

Many of the literary travelers who reported on the newly accessible exotic islands of the Pacific also visited Australia or New Zealand, or both; but almost nothing comes of these visits, and there is hardly any imaginative representation of them by the English between Anthony Trollope's *John Caldigate* (1879) and Lawrence's superb *Kangaroo* (1923). Perhaps this was, in the case of Australia at least, because in these years the country seemed to be too civilized to be a source of romance, and yet without sufficient tradition or separate identity to be culturally interesting. The one exception is the creation of a romantic and heroic outback bushranger in E.W. Hornung's *Stingaree* (1905) and other stories.

It is primarily as a place of emigration that these settlement colonies, or dominions as they should now be called, figure in English writing. In Hardy's *Jude the Obscure* (1895), for instance, Jude's first wife, Arabella, leaves him and goes with her family to Australia, where she commits bigamy; the family doesn't thrive and returns to England, Arabella with them. When she tells Jude about her second husband he is shocked, but she says that there is much greater freedom in sexual matters over there, lots of people do it. Whether Sydney did in fact offer a world freer of conventional social restraints is less important than the fact that Hardy believed it did. In the same novel, Hardy suggests, though he does not explore, another idea about the colonies, when Sue Bridehead says of young Jude (Arabella and Jude's son), known as Little Father Time, that "it is strange . . . that these preturnaturally old boys almost always come from new countries."

South America R.B. Cunninghame Graham was perhaps the most celebrated and vocal advocate for the consideration of the culture of Spanish-speaking South America—indeed of Hispanic culture in general. In a number of volumes (*The Ipané*, 1899; *Thirteen Stories*, 1900; *Success*, 1902; *Progress and Other Sketches*, 1905; *Faith*, 1909; *Hope*, 1910; *Charity*, 1912; and *A Hatchment*, 1913) Cunninghame Graham included the stories and sketches of life across South America from Colombia to Argentina. Many of them are autobiographical or have an autobiographical basis, and he offered English readers a series of unsentimental and very sharply focussed images of a continent he felt had been unjustly neglected in England. The best of his stories demonstrate a deep sympathy for the oppressed, whether women, Indians, or animals, and an equally bitter anger toward their oppressors. Graham was a paradoxical character—an aristocrat who was passionately anti-imperial, aggressively socialist—and though not a major writer, he strongly influenced contemporaries like Bernard Shaw and Conrad; indeed he provided incidents and characters for both men.

W.H. Hudson's most popular novel, *Green Mansions* (1904), is an earth-lover's romance that uses the fertility and vitality of the tropical rainforest, together with its relatively unexplored state, to enable the relationship between the hero and the allegorical birdlike creature (who becomes human) who is the heroine. Hudson grew up in Argentina, and in the episodic *The Purple Land* (1885, revised 1904) he gives half-fictionalized accounts of his youth, particularly on the pampas and in the Argentinian army that fought with Paraguay. This feature of Latin America—as a place of political instability and of almost perpetual civil and interstate warfare—attracted Conrad to write the best-known South American novel of the period, *Nostromo*.

North America There are relatively few imaginative representations of the United States in English literature of the period, considering how many of the most prominent writers visited the country and wrote accounts of their journeys. What is more frequently found is that writers put Americans into an English context.

Kipling married one, and lived for a few years near his wife's family in Vermont. His *Captains Courageous* (1897) is the only novel by a canonical writer to be set entirely in America, and even then most of it takes place at sea, in a fishing vessel on the Grand Banks off New England and Canada. Despite this, the novel embodies much of what Kipling saw as significant about the United States: the boat, which has the symbolic name *We're Here*, is manned by folk from all parts of the world. Harvey, the pampered, effete fifteen-year-old son of a railroad magnate, is flung overboard from a liner on its way to Europe, and is picked up by a dory sent out from the *We're Here*. For several months he is schooled in the egalitarian harshnesses, skills, and pleasures of this microcosmic life. Once ashore, Harvey telegraphs his father in San Diego that he is safe, and there follows a chapter in which the railway millionaire is rushed across the continent at record speed by his own and other companies. The father and son are well pleased with each other; the son has emerged from self-indulgent softness, and the father has discovered what his son can mean to him. The business of civilized (or over-civilized) youth being tested and tempered by struggles with wilderness and the elements is found throughout Kipling's work, but in *Captains Courageous* the ultimately harmonious functioning of the odd mixture of races aboard the *We're Here* illustrates his optimistic vision of the United States, and the extended description of the transcontinental hustle on the railroads represents Kipling's sense of the forces that drive the country—expansion, rapidity, power, and money. Kipling is concerned that there is a threat that the newly affluent country will loose touch with the dangers and privations of the years of its westward expansion, and thus become enfeebled as Harvey is enfeebled at the beginning of the novel. The new understanding between father and son is Kipling's way of showing how important for the strength of the national family are tests like those Harvey has undergone.

In 1879, Stevenson made the journey across the Atlantic and across the continent as *The Amateur Emigrant* (1880). He is a fantasist even in face of the poverty and misery of the professionals; America here and in *The Silverado Squatters* (1883), the record of his honeymoon in California with the woman he had traveled so far for, seems to be primarily a source of outrageous similes and metaphors that reveal with clarity Stevenson's deep concern with himself and with his reputation as a disciple of Meredith.

Meredith himself hardly touches America, but he is clear that the United States offers a substantial economic threat to England. For example, in *The Egoist* (1879) he sends Willoughby Patterne on a world tour after his first jilting, a tour that commences with the United States. There are perhaps two pages of narration, but they are as telling as any two in the novel in contributing to the understanding of Patterne's character. Carrying class conditioning as part of his baggage—"the English gentleman wherever he went"—he writes of the English and the Americans that "where we compared, they were absurd; where we contrasted, they were monstrous," and contrives arrogantly to insult the president. "We go on in our way; they theirs, in the apparent belief that Republicanism operates remarkable changes in human nature." But for all his criticism of their different and plebeian systems and manners, Patterne is not stupid, and his experience led him ever afterwards to speak "respectfully and pensively" of America, for "there are cousins who come to greatness and must be pacified, or they will prove annoying. Heaven forefend a collision between cousins!" (57–59)

Ford's *The Good Soldier* may be taken as an example of a novel that uses an American in a crucial role, in this case the narrator Dowell, through whose consciousness alone we gather whatever we gather about the events in America, Europe, and England. Dowell sees himself as characteristically American, without "the hot passions of these Europeans" (76), as "fainter" (257), regarding "Europe as a sink of iniquity" (89), recording the "aspiration that all American women should one day be sexless" (95). But yet, because he is American, he has a clear perception of the grotesque inadequacies of the English class of "good people" who, taking everything for granted (as Richardson also observes in *Pointed Roofs*), have exported their culture of surface certainty and inner chaos across the world. The kind of cosmopolitanism that the half-German, wholly English Ford brought to the practice of literature is well caught in lines from his poem "To All the Dead," from *High Germany* (1912):

"Crossing the Rocky Mountains," Samuel Manning,
from his *American Pictures* (1877).

63

Do you know the Hudson?
A sort of a Moselle with New York duds on,
There are crags and castles, a distance all grey,
Rocks, forests and elbows.

Almost all the celebrated writers of the period, from Oscar Wilde to Rupert Brooke, visited the United States during this period, and most of them wrote about their experiences. If it is possible to reduce such a heterogeneous body of writing to a generalization, it might be valid to say that they all used America as a looking-glass that would show them more clearly who they were and where they stood in relation to the world. America, both physically and culturally diverse, by turns densely and sparsely inhabited, so advanced and yet so primitive, so vast and yet unified by a single fabric of political structure, offered self-expressive, self-reinforcing matter for any writer.

For a final English view of America and Americans, a little poem called "On a Rhine Steamer" by J.K. Stephen, the ill-starred cousin of Virginia Woolf, might be thought of as representative:

Republic of the West,
  Enlightened, free, sublime,
Unquestionably best
  Production of our time.

The telephone is thine,
  And thine the Pullman Car,
The caucus, the divine
  Intense electric star.

. . . .

But every now and then,
  It cannot be denied,
You breed a kind of men
  Who are not dignified,

Or courteous or refined,
  Benevolent or wise,
Or gifted with a mind
  Beyond the common size,

Or notable for tact,
  Agreeable to me,
Or anything, in fact,
  That people ought to be.[8]

Ireland Ireland was one of the leading political topics of the period; the early years were dominated at Westminster by Parnell's Home Rule party and in Ireland by Michael Davitt's Landleague, about which Trollope wrote his last, unfinished novel *The Landleaguers* (1881), a good deal less sympathetic to the Irish than his earlier works set in Ireland. The strongest English literary (and personal) response came from the radical anti-imperialist Blunt, who was imprisoned for his commitment to the Landleague and subsequently wrote a sonnet sequence, *In Vinculis* (1889), deriving from his experience in jail. After Parnell's death, the Irish MPs were less cohesive as a force under Redmond, but they still intermittently held the balance of power, when they could demand the introduction of Home Rule bills. When the last of these seemed close to success in 1914, there was almost civil war between the Home Rule majority and the Protestant Unionists in Ulster, led by Carson; civil strife was only averted by the declaration of the Great War. In 1916 there was the famous Easter Rising of Nationalists in Dublin, rapidly suppressed by English troops and gunboats; the unwise execution several days later of a number of the leaders ensured that once the Great War was concluded England would face her first serious colonial war of independence since 1776.

This series deliberately excludes Irish writers from consideration, treating Ireland as England's oldest colony; even those Irish men and women of English ancestry belong, as Jonathan Swift well knew, to the tradition of the colonized, not of the colonizers. Nevertheless, it is important to note that during this period for the first time there emerged a group of writers from a colony whose collective voice was significant enough to reach the ears of the wider world.

The Irish Literary Renaissance, as the explosion of talent has come to be known was dominated, and to a degree stage-managed, by William Butler Yeats, but John Millington Synge, George Russell, Augusta Gregory, and George Moore also played significant parts. The movement attracted to it some relatively minor non-Irish writers who saw in the newer Celticism an antidote to English middle-class materialism. The composer Arnold Bax, for instance, wrote several volumes of Irish stories and poems under the pseudonym Dermot O'Byrne. At first the movement was self-consciously concerned with Irish legendary and mythical past; William Sharp, alias Fiona Macleod, who was the leading writer in the contemporary revival of Scottish Celticism, was a strong advocate of pan-Celticism, though firmly within the context of English literature, and he wrote two plays and a number of poems that use Irish material; the most famous of these is "The Immortal Hour" (1907), which gained its widest audience as libretto for Rutland Boughton's opera of the same name.

65

NOTES

Quotations are from English first editions except for the following: Blunt, *My Diaries* (New York: 1923); Woolf, *The Voyage Out* (London: 1965); Conrad, *Heart of Darkness* (in *Youth, Heart of Darkness, The End of the Tether*, London: 1946); Flecker, *Collected Poems* (New York: 1916); Blunt, *Satan Absolved* (in *Collected Poems*, London: 1914); Conrad and Ford, *The Inheritors* (New York: 1924); Sinclair, *The Tree of Heaven* (New York: 1917); Symons, *Poems* (London: 1924); Bennett, *The Old Wives' Tale* (London: 1948); Lewis, *Tarr* (London: 1968); Douglas, *The South Wind* (London: n.d.); Butler, *The Way of All Flesh* (Harmondsworth: 1966); Richardson, *Pointed Roofs* (New York: n.d.); Forster, *Howards End* (London: 1973); Hardy, *Tess of the d'Urbervilles* (Oxford: 1983); Conrad, *Under Western Eyes* (Harmondsworth: 1957); Gissing, *The Crown of Life* (Hassocks, Sussex: 1978); Meredith, *Beauchamp's Career* (London: 1914); Ford, *The Good Soldier* (New York: 1989).

1. James Scully, *Italian Travel Sketches* (London: 1912), 307; quoted in John Pemble, *The Mediterranean Passion* (Oxford: 1987), 27.
2. Hugh Cortazzi and George Webb, eds., *Kipling's Japan: Collected Writings* (London and Atlantic Highlands, NJ: 1988), 35.
3. Ernest Dowson, *The Poems of Ernest Dowson* (London: 1905), 152.
4. Quotations are from an edition published in London in 1977.
5. *Kipling's Japan*, 92.
6. *The Savage South Seas*, painted by Norman H. Hardy (described by E. Way Elkington), 3–4.
7. Quotations are from vol. 4 of *The Novels and Tales of Robert Louis Stevenson* (New York: 1895).
8. Kevin Crossley-Holland, ed., *The Oxford Book of Travel Verse* (Oxford: 1986), 372.

ENGLAND, EUROPE, AND EMPIRE: HARDY, MEREDITH, AND GISSING

Simon Gatrell, University of Georgia

Britain's economic and political primacy among world nations, more or less unquestioned since Trafalgar and Waterloo, came more and more intensely into question during the last quarter of the nineteenth century—after the astonishing successes of the Prussians against Austria and France in 1866 and 1871, the establishment of the new German Empire, and the emergence of the United States from the consequences of the Civil War. There was a gradual erosion in British national self-confidence which led, toward the end of the century, to a steady increase in enthusiasm for Imperial expansion as a way or reestablishing a commanding position among the nations.

At the same time, foreign travel became more and more a matter of course for more and more people; railways had reached most parts of Europe by 1876; travel time to other parts of the world was decreasing rapidly as ships utilized steam power more efficiently; indeed the first recorded use in the *OED* of the description "globe-trotter" is in 1875. It was a time during which no British imaginative writers who were at all interested in the public affairs of their country could avoid feeling some pressure to incorporate aspects of Britain's relations with the world beyond its shores into their work. The work of George Meredith, Thomas Hardy, and George Gissing is both representative of the range of responses to "abroad" and represents the finest of the period's literature.

In several ways it is possible to see Meredith and Gissing on one side of the question of the importance of ideas of abroad to their writing, and Hardy on the other. Meredith and Gissing were both political journalists for a while, whereas Hardy held it to be the duty of a creative writer to refrain from all public comment on political issues; and this difference is clearly present in their fiction. Meredith and Gissing introduce analysis of Britain's contemporary relations with other nations and (in Gissing at least) of her Imperial policy into their novels; Hardy never does. Meredith and Gissing set substantial fragments of some of their most successful novels abroad, and explore through them many of the reasons why people do travel away from Britain, and how what they find responds to their avowed and unavowed aims; the only novels in which Hardy does this at all, *The Hand of Ethelberta* (1875) and *A Laodicean* (1881), are among his least satisfactory or characteristic.

So far as questions of place are concerned, Hardy is, we are accustomed to say, the novelist of Wessex, the writer concerned before all others with recreating in his fiction a corner of England; and yet abroad *is* important even to Hardy. Though his characteristic environment is severely circumscribed, he finds it necessary to make clear to his audience that his characteristic action is universal. Here, for example, is a familiar passage from near the beginning of *The Woodlanders* (1887):

> Hardly anything could be more isolated, or more self-contained, than the lives of these two walking here in the lonely hour before day, when grey shades, material and mental, are so very grey. And yet their lonely courses formed no detached design at all, but were part of the pattern in the great web of human doings then weaving in both hemispheres, from the White Sea to Cape Horn. (24)

Another celebrated extract, this time from the beginning of *The Return of the Native* (1878), shows that on occasion Hardy was also concerned to direct the reader's perception of his half-created environment within a wider than purely local context; he wrote of Egdon Heath:

> Haggard Egdon appealed to a subtler and scarcer instinct, to a more recently learnt emotion, than that which responds to the sort of beauty called charming.
> Indeed, it is a question if the exclusive reign of this orthodox beauty is not approaching its last quarter. The new Vale of Tempe may be a gaunt waste in Thule. . . . And ultimately, to the commonest tourist, spots like Iceland may become what the vineyards and myrtle-gardens of South Europe are to him now; and Heidelberg and Baden be passed unheeded as he hastens from the Alps to the sand-dunes of Scheveningen. (4–5)

The Return of the Native is also unusual in that the native of the title, Clym, is returning from Paris, "that rookery of pomp and vanity," to which he had been driven by the ambitions of his mother; there are hardly any other significant references to Europe in what Hardy called his "novels of character and environment." Escape for Eustacia—the ultimately impossible escape from the social isolation/insulation of Egdon Heath—she hopes at first will be through Clym to Paris; this hope drives all that Eustacia does. And Clym finds that his European-learned ideals of improvement through education are quite impracticable on the heath, and also that they have made it impossible for him to think like a native any longer.

However, for many more characters in Hardy's fiction, major and minor alike, there is a constant economic pressure under the contemporary harsh nature of agricultural employment to find ways of escaping. Thus emigration rather than travel is the characteristic purpose for envisioning abroad in Hardy's novels; abroad is where opportunities seem to present themselves that are denied to those who stay in England. This is true for Clym as it is for Stephen Smith in India, or Gabriel Oak and Donald Farfrae proposing to go to America, or Tim Tangs on his way to New Zealand, or Angel Clare in Brazil, or the Donns in Australia.[1] Some of these have other reasons in addition to economic opportunity for leaving England; some in the end never go, others return disillusioned; the point, though, is that they all face the consequences of leaving, and determine to go, most of them for good. This emigration is a strand in the evidence of the gradual destruction through the century of rural society which it was Hardy's painful pleasure to chronicle; indeed the tendency towards emigration might be seen as one of the defining elements of Wessex.[2]

One of Hardy's works, however, throws a raking light across all this fiction: *The Dynasts* (1903–1908), his massive verse drama of the Napoleonic wars. It is as if the play takes the fragment of *The Woodlanders* quoted above and reinterprets it in a context that can set aside the demands of narrative verisimilitude. In the mimetic frame of the novel we understand the network of human doings across the world as merely an expansion of the tiny patch of the web that Hardy focuses on; the perspective remains essentially a human one; but in *The Dynasts* we are continually being shown human events from a point of view so remotely distanced from the earth that the inconsequence of whole armies, let alone of individual actions, becomes our habitual thought. Thus the play's second stage direction:

> The nether sky opens, and Europe is disclosed as a prone and emaciated figure, the Alps shaping like a backbone, and the branching mountain-chains like ribs, the peninsular plateau of Spain forming a head. Broad and lengthy lowlands stretch from the north of France across Russia like a grey-green garment hemmed by the Ural mountains and the glistening Arctic Ocean.

The point of view then sinks downwards through space, and draws near to the surface of the perturbed countries, where the peoples, distressed by events which they did not cause, are seen writhing, crawling, heaving, and vibrating in their various cities and nationalities. (6)

At the end of the second part of the drama, one of the disembodied commentators on the action, Chorus of the Years, invokes Europe in order to show how even the trees and the oceans are moved by the unconscious amoral force of the Immanent Will, here called the "rapt Determinator":

> Why watch we here? Look all around
> Where Europe spreads her crinkled ground,
> From Osmanlee to Hekla's mound,
>   Look all around!

> Hark at the cloud-combed Ural pines;
> See how each, wailful-wise, inclines;
> Mark the mist's layrinthine lines;

> Behold the tumbling Biscay Bay;
> The Midland main in silent sway:
> As urged to move them, so move they.

> No less through regal puppet-shows
> The rapt Determinator throes,
> That neither good nor evil knows!

CHORUS OF THE PITIES

> Yet It may wake and understand
> Ere earth unshape, know all things, and
> With knowledge use a painless hand,
>   A painless hand! (322)

This reminder of the inconsequence of human actions, and of Hardy's hope that the Immanent Will may be evolving benificent consciousness, is only one of many throughout the play. But it is through his stage directions that Hardy instructs us most clearly in the transformation of scale that his vision demands; this, for instance, is how we are asked to envisage the battle of Leipzig:

So massive is the contest that we soon fail to individualize the combatants as beings, and can only observe them as amorphous drifts, clouds and waves of conscious atoms, surging and rolling together; can only particu-larize them by race, tribe, and language. Nationalities from the uttermost parts of Asia here meet those from the Atlantic edge of Europe for the first and last time. By noon the sound becomes a loud droning, uninterrupted and breve-like, as the sound from the pedal of an organ kept continuously down. (383)

Hardy's sense of the smallness of the world, its interconnectedness and interdependence, of the futility of nationalistic or dynastic ambitions in the face of this perspective, is really a very modern one. Hardy has no sympathy for such dynastic warfare, and in this massive drama as much as in the brief famous lyric "In Time of 'The Breaking of Nations,'" he balances against it the activities and values of ordinary lives. All human endeavor, when seen from so remote a controlling perspective as that in *The Dynasts*, seems futile, but poised against such an account is Hardy's vivid representation of the individual combatant, who, whether Captain Hardy supporting the dying Nelson or Hussar Sergeant Young at bivouac on the eve of the battle of Vitoria, recalls with deep affection the life of the home from which war has driven him. The remote but total control that is exercised over the action by the Immanent Will renders the significance of the local and the individual more powerful, but also more poignant.

Going for a soldier was, of course, another common resource for the displaced rural workfolk; Hardy remembers Drummer Hodge killed in South Africa during the Boer War, who "never knew— / Fresh from his Wessex home— / The meaning of the broad Karoo," and in this poem, too, Hardy intensifies the poignancy of the soldier's experience of the other, both through death abroad (as he does with those who died on the Isle of Walcheren in *The Dynasts*) and through the impersonality of the forces that take him to the other side of the world. "Yet portion of that unknown plain / Will Hodge for ever be": you carry with you what you are, and where you come from, wherever you go, wherever you die. It is a simple, even a self-evident conclusion, but one that lies also at the heart of Meredith's accounts of those who travel, as it does of Gissing's; perhaps indeed it is simply a universal truth. It is also most often the case in the work of these three novelists that anyone who chooses to travel abroad from England is traveling to escape something; there are very few who travel for education, to widen their horizons, for refreshment, or even really for new sensations, except insofar as the new sensations are meant to efface others.

In contrast with Hardy, Meredith is a novelist of the affluent classes—the aristocracy, the landed gentry, owners of mines and successful stock-brokers—and travel abroad is an integral part of their lives. But it is still true

Poster advertising Granville Barker's production of Hardy's
The Dynasts (1914), Charles Ricketts.

that in his novels the reason most major characters travel is to escape or to forget.

The majority travel to forget unhappy love.[3] Harry Richmond climbs in the Alps and bakes in the Egyptian desert in order to forget Princess Ottilia, and goes on an Indian expedition in order to forget Janet Ilchester; Cecilia . Halkett travels to Italy to forget Nevil Beauchamp and to escape her jealousy of Renée de Croisnel, her rival for his love. Sometimes this unhappiness is mixed with potential social consequences; thus Willoughby Patterne goes on a world tour to efface the unfortunate public impression created by his being jilted by Constantia Durham, and Diana Warwick goes voyaging in the Mediterranean and wandering in the Alps to escape the aftermath of her husband's attempts to divorce her. Sometimes in this gradation it is the social aspects that constitute the primary force driving the character abroad: thus Nataly Dreighton, unmarried but living with Victor Radnor, is taken to France and to Italy to try to diminish the emotional and moral pressure Radnor has put her under by continually thrusting her into situations of great social publicity. Some travel to escape financial as well as social problems in England: Richmond Roy goes to Germany and elsewhere at various times to escape his creditors and social scandal; others travel to escape domestic anxieties, like Nesta Victoria Radnor, who goes to France to escape the terrible situation after the death of her mother.

What they all find is, however, not the same thing. Richmond Roy, time after time, manages to refurbish his reputation and fortune by a period of activity abroad, most vividly and representatively embodied in his service as a glorified clown-prince to the court at Sarkeld; but it is relatively easy to reap the desired reward from foreign travel when you are not looking for a transformation in your consciousness. This is true also in a different measure for Diana Warwick, who finds in the mountains above Lugano: "Freedom to breathe, gaze, climb, grow with the grasses, fly with the clouds, to muse, to sing, to be an unclaimed self, dispersed upon earth, air, sky, to find a keener transfigured self in that radiation" (*Diana of the Crossways* 139). The renewal of the simplicity, innocence, and selfhood that Meredith associates with the Alps is permanent in Diana because, though she has been desperately unhappy in her choice of husband, her heart is whole, and she is able genuinely to open herself to these impressions.

Most of those who are running from themselves are at best only partially successful; many fail completely to escape or forget, for the first law of travel for these writers is that though you may easily change your location, it requires a powerful combination of self-knowledge, willpower, and openness to fresh impressions to efface a deeply engraved emotion from your mind. Of course the success of the effacement does depend upon the depth of the engraving. Patterne's willpower is quite sufficient, without much self-knowledge, to refurbish his self-esteem (though the word *jilt* remains a sensitive spot in his consciousness). On the other hand, though Meredith writes in *One of Our Conquerors* that "travelling shook Nataly [Dreighton] out of her troubles and gave her something of the child's inheritance of the wisdom of life—*the living ever so little ahead of ourselves*," he added sardonically "about as far as the fox in view of the hunt"(149). And though she knows she should live in the hour, though "escape and beauty beckoned ahead" she cannot for long forget the "chains" that are behind her in England. Her mind is almost literally torn apart between her love for Victor, her anxiety for the social position of her daughter, and her terror of the nakedness of the public prominence into which he has thrust them both. No exposure to unfamiliar cultures or spectacular scenery can efface such deep-seated anxiety and terror.

Harry Richmond, however, does find relief from his loss of Princess Ottilia, and from his consequent loss of purpose in life, in confronting the Alps:

> Carry your fever to the Alps, you of minds diseased: not to sit down in sight of them ruminating, for bodily ease and comfort will trick the soul and set you measuring our lean humanity against yonder sublime and infinite; but mount, rack the limbs, wrestle it out among the peaks; taste the danger, sweat, earn rest: learn to discover ungrudgingly that haggard fatigue is the fair vision you have run to earth, and that rest is your uttermost reward. Would you know what it is to hope again, and have all your hopes at hand?—hang upon the crags at a gradient that makes your next step a debate between the thing you are and the thing you may become. (*The Adventures of Harry Richmond* 539–40)

Though this physical assault of the mountains renews some of the springs of ordinary life in him, he has a deeper unconscious need that danger and exhaustion cannot reach; he flies to the other extreme and tries oblivion in the heat of the desert, only to find that a dream of rain and home drives him to England and Janet Ilchester.

In *Beauchamp's Career*, Cecilia Halkett's reasons for choosing Italy as the goal of her journey to forgetfulness seem to look back to an experience similar to Diana Warwick's above Lugano: "It was in Italy that Cecilia's maiden dreams of life had opened. She hoped to recover them in Italy, and the calm security of a mind untainted. Italy was to be her reviving air"(527). But during the journey across France Nevil Beauchamp haunts Cecilia's dreams; and her arrival in Italy cannot satisfy her:

I am in Italy! she sighed with rapture. The wine of delight and oblivion was at her lips.

But thirst is not enjoyment, and a satiated thirst that we insist on over-satisfying to drown the recollection of past anguish, is baneful to the soul. In Rome Cecilia's vision of her track to Rome was of a run of fire over a heath. She could scarcely feel common pleasure in Rome. It seemed burnt out. (527–28)

Her pilgrimage is fruitless; she cannot forget Beauchamp even in Italy, and her observation of less cultivated, more pretentious English abroad is couched in terms borrowed from the fiercely radical mind that continues its hold over her:

An ultra-English family in Rome, composed, shocking to relate, of a baronet banker and his wife, two faint-faced girls, and a young gentleman of our country . . . chose to be followed by their footman in the me-lancholy pomp of state livery. . . .

". . . Those English sow contempt of us all over Europe. We cannot but be despised. One comes abroad foredoomed to share the sentiment. This is your middle-class! What society can they move in, that sanctions a vulgarity so perplexing? They have the air of ornaments on a cottager's parlour mantelpiece. . . . We scoff at the vanity of the French, but it is a graceful vanity; pardonable, compared with ours." (528–29)

Here Meredith is continuing a long tradition of disgust expressed by English writers of all periods at the vulgar English abroad, the English who appear to be traveling in order to reaffirm their superiority to the peoples amongst whom they travel. There is another devastating example of the species at the beginning of *Vittoria* (1867), his novel of the Italian struggle for independence. This time it is seen through the eyes not of a superior English visitor, but of the native inhabitants of Italy:

Carlo Ammiani had descried the advanced troop of a procession of gravely-heated climbers—ladies upon donkeys, and pedestrian guards stalking beside them, with courier, and lacqueys, and baskets of provi-sions, all bearing the stamp of pilgrims from the great Western Island. . . .

A mountain ascended by these children of the forcible Isle, is a moun-tain to be captured, and colonized, and absolutely occupied for a term; so that Vittoria soon found herself and her small body of adherents observed, and even exclaimed against, as a sort of intruding aborigines, whose presence entirely dispelled the sense of romantic dominion which a mighty eminence should give, and which Britons expect when they have expended a portion of their energies. (50)

This description is particularly good, in that it gives a metaphor for British attitudes to power and Empire, as well as to the tourist centers of Europe, though even in his later fiction Meredith was hardly interested in Imperialism—his main perception concerning the expansion of the colonies was that economic competition from Europe was the force behind the expansion; unless Britain could grow economically, Germany or France would slowly starve it out of its old markets and securities; this is particularly clear in *One of Our Conquerors*.

Meredith was one of the most European of Victorian writers; he was partly educated in Germany; his second wife was half French; he was almost as fanatical a supporter of the cause of Italian unification and liberty as Swinburne, an admirer of Germany and a lover of France. He defined himself at the time of the Franco-Prussian War in a letter of February 27, 1871: "I am neither German nor French, nor, unless the nation is attacked, English. I am European and Cosmopolitan—for humanity! The nation which shows most worth, is the nation I love and reverence"(1.440–1). As a writer, the question of relationships between England and abroad, particularly Europe, is one of those at the center of his thought, one of a handful of controlling ideas that surfaces again and again in his fiction and poetry in different guises. As a traveler abroad himself, it was first of all the differences in the earth itself, in the landscape, that moved him. His first experience of the Alps entered his spirit as deeply as it did any of the Romantic poets. He wrote in a long letter to Maxse, dated July 26, 1861:

> My first sight of the Alps has raised odd feelings. Here at last seems something more than earth, and visible, if not tangible. They have the whiteness, the silence, the beauty and mystery of thoughts seldom unveiled within us, but which conquer Earth when once they are. . . . Our great error has been . . . to raise a spiritual system in antagonism to Nature. (1.93)

For Gissing, on the other hand, it was the remains of classical Rome and Naples that moved him most powerfully on his long-delayed first journey to Europe in 1888–89. Gissing's novels most often grew directly from his immediate personal experience, and all critics agree that *The Emancipated* (1890), the novel he wrote as a result of this trip, planned in Venice before he returned to England, has quite a different feel from his earlier work. It is the first time that the world beyond England entered with any significance into the fabric of his work, and that broader perspective remains through the rest of his writing, though the only other completed novel that has much action set away from England is *Sleeping Fires* (1895), a relatively small part of which was suggested by his stay in Athens in 1889. He excluded almost entirely from his later novels the very poor and the lower classes, who on the whole didn't go abroad, and had no compelling interest in it—though Gissing himself, as his diaries show, had to watch every penny, and was continually casting about for somewhere cheaper to eat or to stay. We do see in *The Emancipated* that some people are abroad for the perennial Victorian reason that it is cheaper to live there: there is Musselwhite, the impoverished scion of a noble house exiled from his home; he

> wandered in melancholy, year after year, round a circle of continental resorts, never seeking relief in dissipation, never discovering a rational pursuit, imagining to himself that he atoned for the disreputable past in keeping far from the track of his distinguished relatives.(43)

Gissing gives a brief account in Denyer of an unsuccessful itinerant international entrepreneur, whose family, currently in Naples, is never sure where he is or where the next meal is coming from—a character from the underside of the great commercial enterprises of the last quarter of the century, the kind of man that Conrad would put at the center of some of his fiction.[4] But the central characters of *The Emancipated* are well-enough off, partly because most travelers were (Gissing could be wittily cynical about them: "the race of modern nomads, those curious beings who are reviving an early stage of civilization as an ingenious expedient for employing money and time which they had not intelligence enough to spend in a settled habitat" [45]), but also because Gissing in most of his later fiction is interested in the fraction of the middle class in which social change seems to be possible.

It is essentially his suggestion, begun in *The Emancipated* and carried on in his later fiction, that mid-Victorian society was so severely structured because it feared that is was *not* thoroughly civilized, that by the end of the 1880s some segments of society were beginning to break out of these structures, to feel no longer the need for the strict formal restraints on behavior. Some ordinary respectable intelligent people began to feel it was alright to be hostile to the prevailing conventions, and Gissing became aware from his own observation that it was easier for Englishmen and, particularly, Englishwomen to set aside established modes of thought and behavior when living abroad. Thus Cecily, Reuben Elgar, Miriam, and Mrs Lessingham, all of whom are (or become) to some degree free from the constraints of English society, learn their new attitudes and ideas abroad.

Genuine emancipation, however, presumes education, sensibility, and self-knowledge—a high degree of civilization; Gissing's narrative demonstrates, in particular, two consequences of living in defiance of conventions. If these qualities are lacking, as, for instance, most of them are in Elgar, then the emancipation will be no more than a mask over moral dissipation. Even possessing these qualities, as by and large Cecily does, unless restraint is added to them, the pioneer will face even intolerable harshness from the still-enslaved majority in English society:

> Life is so simple to people of the old civilization. The rules are laid down so broadly and plainly, and the conscience they have created answers so readily when appealed to. But for these poor instructed persons, what a complex affair has morality become! Hard enough for men, but for women desperate indeed. Each must be her own casuist, and without any criterion save what she can establish by her own experience. The growth of Cecily's mind had removed her further and further from simplicity of thought. . . . Her safeguard was an innate nobleness of spirit. But it is not to every woman of brains that this is granted. (423–24)

Gissing develops early on and sustains throughout an almost violent contrast between the culture, society, religion, climate, and topography of southern Italy and northern Britain. The novel begins with a picture of Miriam Baske "by a window looking from Posillipo upon the Bay of Naples . . . her dress . . . was severely plain, and its grey coldness, which would well have harmonized with an English sky in the month of November, looked alien in the southern sunlight" (3). Gissing uses Naples and Italy in general, in all its aspects, as a correlative for genuine emancipation, and Manchester, and industrial-commercial England, in all its aspects, as a correlative for slavery.[5]

Miriam, a puritan from northern industrial England, is like an extreme version of George Eliot's Dorothea Brooke, more or less abandoned in Italy without anything in her past that would give her a way of understanding what she is experiencing; but unlike Dorothea she is given time and opportunity to learn what Naples and Rome can mean. In company with the painter Ross Mallard she begins to respond to southern landscape, Italian literature, Pagan and Catholic art and architecture; she discovers herself behind the grimy cracking façade of doctrine, and gradually sheds the constricting shell of her cultural and religious inheritance.

Mallard is also from the north of England, and the contrast between north and south is carried on through him as well. To Miriam he makes a distinction between the work and the life later made memorable in a poem by Yeats:

"I can do better work when I take subjects in wild scenery and stern climates, but when my thoughts go out for pleasure, they choose Italy. I don't enjoy myself in the Hebrides or Norway, but what powers I have are all brought out there. Here I am not disposed to work. I want to live, and I feel that life can be a satisfaction in itself without labour." (334)

Miriam's Italian experience has enabled her to free herself from the narrow limitations of her earlier life, but her emancipation is tempered by a respect for what seems reasonable in Victorian social conventions. Gissing's representation at the end of the novel of her marriage to Mallard shows this moderate conservative thoughtfulness.

The Whirlpool (1897) is also in part a novel concerned with emancipation, and by this time Gissing's attitude to it has hardened; he implies that perhaps the mid-Victorians were right; that mankind had emerged not so far from savagery as some had thought. He shows on the one hand, in the figure of Sibyl Carnaby, that behind the most thoroughly emancipated façade may be a brutal self-serving nature. On the other hand, in the character of Alma Rolfe, he shows that freedom is destructive without self-knowledge and strength of intellect. Gissing suggests that Alma would have been safer, happier—still alive, perhaps—if Rolfe had been a conventional middle-class Victorian husband.

The discussion of this issue in The Whirlpool modulates into an account of a surge in nationalistic Imperialism in all classes of English society; it had become acceptable for ordinarily cultivated, peaceable, middle-class people to respond fervently to the idea of Empire. In place of the highly educated, sensitive, intelligent rejection of constricting Victorian society proposed in The Emancipated, Imperialism in The Whirlpool offers a return to a healthy savagery. Harvey Rolfe says enthusiastically to his friend Hugh Carnaby at the beginning of the novel:

"I was looking at a map in Stanford's window the other day, and it amused me. Who believes for a moment that England will remain satisfied with bits here and there? We have to swallow the whole, of course. We shall go on fighting and annexing, until—until the decline and fall of the British Empire. That hasn't begun yet. Some of us are so over-civilised that it makes a reaction of wholesome barbarism in the rest." (16)

In Hugh Carnaby we are shown a representative new barbarian, a man who is no explorer, but whose only comfort is in strenuously experiencing (mostly with a gun) remote wild places, who is intolerably reined in by the suffocating limitations of England; a man who can say of himself: "I'm a fierce, strong brute, who ought to be anywhere but among civilized people."

Like so many of Meredith's characters, he travels to escape, though with indifferent success. It is one of the controlling ironies of the novel that Carnaby's wife, Sibyl, who hates the names of the new colonies because they imply barbarism, who is offered as the epitome of emancipated Victorian civilization, is shown to be by far the more brutal of the two.

At the end of the story, after he has married, had a son, and seen his wife kill herself, Rolfe reads Kipling's *Barrack-Room Ballads*, and comments:

> The Empire; that's beginning to mean something. The average Englander has never grasped the fact that there was such a thing as a British Empire. He's beginning to learn it, and itches to kick somebody, to prove his Imperialism. The bully of the music-hall shouting "Jingo" had his special audience. Now comes a man of genius, and decent folk don't feel ashamed to listen this time. We begin to feel our position. We can't make money quite so easily as we used to; scoundrels in Germany and elsewhere have dared to learn the trick of commerce. We feel sore, and it's a great relief to have our advantages pointed out to us. By God! we are the British Empire, and we'll just show 'em what *that* means!" (450)

Whether we accept H.G. Wells's view that Rolfe was here robustly expressing his own opinion, or Gissing's claim in response to Wells that, having matured through experience, he speaks with quiet sarcasm, either way the analysis of a new mood among "decent men" both articulated and stimulated by the writing of a great artist is striking. Part of Rolfe's response to Kipling is a clear-sighted prediction to his friend Morton of the Great War, in which the enthusiasm certainly does seem ironic:

> It's a long time since the end of the Napoleonic wars. Since then Europe has seen only the sputterings of temper. Mankind won't stand it much longer, this encroachment of the humane spirit. . . . We may reasonably hope, old man, to see our boys blown into small bits by the explosive that hasn't got its name yet. (449–50)

Gissing's own view seems clear from the admirable Morton's response to this speech: "I'm reading the campaigns of Belisarius" (Belisarius was a commander during the final defeat of the Roman Empire).

The two years between the writing of *The Whirlpool* and *The Crown of Life* (1899) brought England to the brink of the Boer War, and though the latter novel is primarily a love story, there is twined about the romance Gissing's further consideration of Imperialism. Irene Derwent is one of Rolfe's cultivated decent people who is moved by a charismatic advocate of Empire (not in this instance Kipling, but a shadowy Milner/Rhodes figure called Trafford Romaine) to advocate Imperialism red in tooth and claw. If Imperialism is not totally discredited in the novel, it is because love and peace are seen as more powerful, and because Gissing, with the Roman example so constantly before him, would have found it hard to reject outright the idea of Empire. It is the way in which it is pursued that he hates, the commercial greed supported by unjustifiable war. A sane and moderate character in the novel (John Jacks) says this:

> "Our pride has been a good thing, on the whole. Whether it will still be, now that it's so largely the pride of riches, let him say who is alive fifty years hence. . . . We're beginning, now, to gamble for slices of the world. We're getting base, too, in our grovelling before the millionaire. . . . Our pride, if we don't look out, will turn to bluffing and bullying. I'm afraid we govern selfishly where we've conquered. We hear dark things of India, and worse of Africa. And hear the roaring of the Jingoes! Johnson defined Patriotism, you know, as the last refuge of a scoundrel; it looks as if it might presently be the last refuge of a fool." (179–80)

It is perhaps not surprising that Gissing's last novel *Veranilda* (incomplete at his death and published posthumously in 1904), is enacted, like Belisarius's campaigns, at the time of the final overthrow of Rome.

Unlike Meredith and Hardy, Gissing wrote a travel book, *By the Ionian Sea* (1901), which is arguably one of his most successful works. As the account of a journey in search of the sites of Magna Graecia along the southern shore of Calabria, it is the reflection of an enthusiastic, innocent, and ironic mind, but it is also, in its tone of gentle regret for the utter decay of what he valued, the account of yet another inevitably indeterminate attempt to escape from the conditions of Victorian England.

NOTES

Quotations from the works of Hardy, Meredith, and Gissing are from the following editions: *The Woodlanders*, ed. Dale Kramer (Oxford: 1981); *The Return of the Native*, ed. Simon Gatrell (Oxford: 1990); *The Whirlpool*, ed. Patrick Parrinder (Rutherford, Madison, Teaneck, NJ: 1977); *The Dynasts* (London: 1919); "Drummer Hodge" in *Collected Poems of Thomas Hardy* (London: 1919); *Diana of the Crossways* (London: 1886); *One of Our Conquerors*, ed. Margaret Harris (St. Lucia: 1975); *The Adventures of Harry Richmond* (London: 1905); *Beauchamp's Career* (London: 1914); *Vittoria* (London: 1914): *The Letters of George Meredith*, ed. C.L. Cline (Oxford: 1970); *The Emancipated*, ed. Pierre Coustillas (Hassocks, Sussex: 1977); *The Crown of Life*, ed. Michael Ballard (Hassocks, Sussex: 1978).

1. In *A Pair of Blue Eyes* (1873), *Far From the Madding Crowd* (1874), *The Mayor of Casterbridge* (1886), *The Woodlanders* (1887), *Tess of the d'Urbervilles* (1891) and *Jude the Obscure* (1895), respectively.
2. It is of passing interest that Rolfe, in Gissing's *The Whirlpool* (1897) once worked in the office of an emigration agent:

   it became one of his functions to answer persons who visited the office for information as to the climatic features of this or that new country, and their physical fitness for going out as colonists. Of course there was demanded of him a radical unscrupulousness, and often he proved equal to the occasion. (22–23)

   A link between Angel Clare and Conrad's Marlow in *Heart of Darkness* (1902)—London seems to Gissing at least to have been as much a whited sepulchre as Brussels.
3. The characters discussed in this section are from the following novels: Harry Richmond and Richmond Roy—*The Adventures of Harry Richmond* (1871); Cecilia Halkett—*Beauchamp's Career* (1876); Willoughby Patterne—*The Egoist* (1879); Diana Warwick—*Diana of the Crossways* (1885); Nataly Dreighton and Nesta Victoria Radnor—*One of Our Conquerors* (1891).
4. There is a strong sense in the later Gissing of an international economy in which an Englishman can hope to turn a penny anywhere, like the man Mackintosh in *The Whirlpool*, who "has been everywhere and done everything—not long ago was in the service of the Indo-European Telegraph Company at Tehran, and afterwards lived at Bagdad, where he got a *date-boil*, which marks his face and testifies to his veracity. He has been trying to start a timber business [in Tasmania]" or Piers Otway in Russia in *The Crown of Life*.
5. More tentatively one might suggest that Paris, though never a scene of action in the novel, is the correlative of false emancipation.

A KIND OF BEAUTY: WILLIAM MORRIS'S TRAVELS TO ICELAND

P.M. Tilling, The University of Ulster

The intense curiosity of the English Victorian for information on foreign and exotic languages, literatures, and places encompassed Iceland and its medieval literature, and something of a "Nordic cult" developed. Throughout the nineteenth century, there grew an informed and scholarly appreciation of early Icelandic (Old Norse) and its saga literature and poetry. As grammars and textbooks of Old Norse became available, writers were able to turn to the original texts, rather than rely on translations from the Latin as, for example, Thomas Gray had earlier done in his poetic "imitations" from the Icelandic. Reliable translations into English of both prose and poetry were also published for the first time, principally by the scholars Sir George Dasent, Samuel Laing, and Benjamin Thorpe. A number of writers, such as Arnold in his poem "Balder Dead" (1855) or Carlyle in his essay on Odin or his *Early Kings of Norway* (1841), were drawn to "that strange island Iceland . . . with its snow jokuls, roaring geysers, sulphur-pools and horrid volcanic chasms, like the waste chaotic battlefield of Frost and Fire."[1] Few who drew upon Icelandic literary materials for their inspiration chose, however, to visit Iceland, and most were content to remodel them to suit their own views and preconceptions.

The new translations enabled a reader to assess medieval Icelandic literature more or less objectively, rather than through the highly individual interpretations of Arnold or Carlyle, and there was, paralleling the interest in the language and literature of early Iceland, a growing interest in Iceland itself. A number of Victorian travelers made the difficult sea journey and several of them wrote about their experiences. Though some were drawn to Iceland because of its literature (Sabine Baring-Gould, for instance, whose

Iceland: Its Scenes and Sagas was published in 1873), others were interested in its geology or were simply tourists, though often they were exceptionally well informed and their published accounts were of considerable literary merit. Thus, Lord Dufferin's *Letters from High Latitudes* (1856) shows a lively and educated interest in various aspects of Iceland: its life, literature, natural history, and geology.[2] Prominent among the Victorian enthusiasts for Iceland and its early literature was William Morris. His work embraces several of the fields covered by others; he was a scholar-translator, author of Icelandic-inspired poems and prose works, and a travel writer. Morris visited Iceland twice (in 1871 and 1873) and both Iceland and its literature helped shape his own writing and, to some extent, his outlook on life.

Morris's first real introduction to Icelandic saga literature was during his search for stories for his early compilation of poems *The Earthly Paradise* (1868–70). His poetic narrative "The Lovers of Gudrun" in *The Earthly Paradise* is based on *Laxdale Saga*. Thereafter, almost all his work owes something to Iceland. That Morris should be so receptive to the literature of medieval Iceland is scarcely surprising since, from an early age, he had shared the Victorian antiquarian enthusiasms, looking to past cultures and literatures in part as a reaction against the sordidness of the industrial present. Though always interested in classical literature, it was the medieval that most captured his imagination, and in the medieval literature of Iceland particularly, he found a toughness, an independence of spirit, and a profound fatalism, with all of which he felt a special sympathy.

Crucial to the development of Morris's Icelandic interests was his meeting in 1868 with the Icelander Eiríkr Magnússon, who was to become librarian of the University Library, Cambridge. They became close friends and, under his tutorship, Morris learned Icelandic with the thoroughness and enthusiasm that characterized all his varied activities. According to Magnússon, Morris "in the lapse of three months mastered the language in a marvellous degree."[3] In fact, without Magnússon, Morris would hardly have been able to translate the sagas; these translations constitute an important part of his literary achievement. Without the close study of text that translation requires, Morris's knowledge of the sagas would have been less complete and his ability to recreate something of their spirit in *The Story of Sigurd the Volsung and the Fall of the Niblungs* (1876), for example, would not have been possible. Magnússon provided him with a literal translation, which Morris then rewrote in a kind of English which, in its use of archaisms and Germanic rather than French-derived forms, recalled something of the style and feel of the original. It was not received kindly in all quarters.[4] *The Story of Grettir the Strong* (*Grettis Saga*) was the first of the translations to be published under the joint names of Magnússon and Morris (1869), to be followed by *The Story of the Volsungs and Niblungs* (*Volsunga Saga*) in 1870

and various of the short tales, which were later (1875) included in the volume
Three Northern Love Stories and Other Tales.[5]

It was inevitable that Morris would wish to travel to Iceland to experience
the landscape of the sagas, to see at first hand the sites of the major saga
events and to attempt to discover something of the spirit that had given them
life. The preparations for his first visit in 1871 are documented in his letters
(137, 139). The party was to consist of four persons: Morris, Magnússon,
C.J. Faulkner (a partner in the firm Morris, Marshall, Faulkner, and Co.)
and a recent acquaintance, W.H. Evans. The group set sail for Iceland (via
the Faroe Islands) on Sunday, July 9, 1871, from Granton, near Edinburgh,
in the Danish mail-boat *Diana*. That the journey should have taken place in
1871 was likely to have been in part the result of turmoil in Morris's private
life, caused by the relationship between his wife Jane and his long-standing
friend Dante Gabriel Rossetti—though, perversely, Morris left them
together in Kelmscott Manor, of which he and Rossetti had recently taken
joint tenancy. Iceland was to serve as a retreat, a place in which he could
perhaps solve his problems by escaping from them. There is, however, no
mention of this in the letters that he wrote from Iceland.

Once in Iceland, Morris was able to experience for himself something of
the rigorous life of the Icelander in his constant battle against weather and
terrain. Morris took on the role of both pilgrim and writer. He visited the
major saga sites and each night wrote an account of the day's travels in a
diary, which he later revised as a *Journal*. He was to do the same when he
returned to Iceland in 1873 and both *Journal* and the *Diary* of 1873 chart in
considerable detail his experiences and impressions of Iceland. Although
these are personal records, they are by no means private, and although
neither was published in Morris's lifetime, it is clear that they were intended
for a wider audience; however, the reader of both *Journal* and *Diary* is
handicapped by the fact that we have them in an unpolished state. Of the
two, the *Journal* is the most finished, for the process of revision and
annotation is virtually complete: at first, the plan was to divide the work into
chapters and within these to separate each day's journey into subsections;
however, after Chapter 5, Morris dispenses with chapter divisions and his
annotations become less frequent. This may have been because he had
abandoned plans for publication or because of other pressures, for he
traveled to Italy in 1872 and to Iceland again in 1873 and, in fact, the final
entries in the *Journal* show that rewriting was still in progress shortly before
he left for Iceland a second time. In any case, Morris states in a letter to
Louisa MacDonald Baldwin that he found it difficult to summon up enthusiasm for the *Journal* after returning to England and regarded it as "but a poor
specimen of its class" (*Letters* 149). However, he did read parts of it to friends,
one of whom, Edmund Gosse, reports of the experience in January 1872:

[Morris] was good enough to say that he had heard all about me and would read the journal because he knew I was interested in Northern matters. As this journal, in spite of statements which have appeared in the papers, is not to be published, it was a great privilege to listen to it. It was very vivid and amusing.[6]

The *Diary*, the record of Morris's second trip in 1873 (on which he was accompanied by C.J. Faulkner) follows the day-by-day plan of the *Journal*, but was not revised by Morris, nor does it cover the entire journey. It commences with the sighting of Iceland and breaks down into a series of notes as the travelers reach Skagafjorð, the most northerly point of their journey. Morris even omits an episode which had greatly impressed and amused him at the time; a fellow traveler on the outward journey, an elderly lady ("an ungrateful & stupid old creature as ever came out of Somerset-shire"), caused considerable interest in Reykjavík, as Morris describes in a letter to Louisa Baldwin ("the old lady wanted to be Guy Faukesed about Iceland in a chair" [*Letters* 199]).

The *Journal* and *Diary* were not published until 1911, when they appeared as Volume Eight of the *Collected Works*.[7] Despite their unpolished state, both *Journal* and *Diary* are of more than documentary interest, containing as they do some of his finest writing (as well as some of his least interesting), and are like no other of his work. Both are more or less straightforward accounts of Morris's day-to-day travels and are often brutally factual, yet they are often lively and often powerfully evocative in their ability to conjure up the splendors and horrors of the Icelandic landscape. Rather than a complete picture of contemporary Iceland, we are shown the country as Morris the writer steeped in saga-lore saw it.

The style of the *Journal* and the greater part of the *Diary* is plain and straightforward, as Morris gives a detailed account of each stage of his journey. The route is recorded with precision and all obstacles and visible topographical features are noted, together with Morris's impressions. He is frequently overpowered by the awfulness and awesomeness of it all, and words like "horrible," "dreadful," "terrible," "frightful," and "black" feature high in his choice of vocabulary. Yet, he is often able to see beauty in the desolation: "Ah, what an awful place! but so barren and dreadful it looks and yet it has a kind of beauty about it" (*Diary* 203); and, again: "Once or twice I looked back at the valley we had left and saw it swept across with mingled rain and sun too, it looked a great hollow far below us soon: a wonderful sight with those terrible mountains at its head" (*Journal* 128).

Frequently, though, one suspects that Morris is happier with a different, more pastoral kind of landscape—one, perhaps, that reminds him of the English countryside. He seizes with delight on the occasional patch of green and the lush meadows surrounding many of the homesteads: "Still, here as in many places, there was a charm about the green sloping meadow and little bright stream running through it, that one could scarcely imagine it could be attained by such simple means" (*Journal* 82).

Morris also delights in identifying the flowers that he comes across, recalling the interest in plant life that provided him with such a fruitful source of inspiration for his wallpaper designs. Despite his obvious sympathies for the land of the sagas, Morris is here essentially the English Victorian abroad. He and his party (despite the presence of Magnússon) behave throughout as English gentlemen, with their shooting and fishing and games of whist in the evenings. As Jan Morris has remarked of the expedition: "The atmosphere was boyish and breezy—public school, upper middle class."[8]

Morris, however, was not in Iceland simply to experience the landscape for its own sake. His first visit, at least, had a more elevated purpose. He wished to experience the major saga sites for himself and, in so doing, understand more clearly the sagas and the force that had given rise to them, for the Icelandic sagas, as he was aware, derive much of their individuality from the landscapes within which they are set. In this respect the journey of 1871 is an example of Morris's perennial wish to become the complete master of any subject that he explored.

In Markfleet, the territory of *Njal's Saga*, Morris experiences at firsthand the source of the saga's inspiration:

> Below was the flat black plain space of the valley, and all about it every kind of distortion and disruption, and the labyrinth of the furious brimstone-laden Markfleet winding amidst it lay between us and anything like smoothness: surely it was what "I had come for to see," yet with that came a feeling of exaltation too, and I seemed to understand how people under all disadvantages should find their imagination kindle amid such scenes. (*Journal* 54)

Occasionally, too, the sight of the actual location of a saga that he knows well causes Morris to see that saga in a new light. On two occasions he records that Iceland has given him new insights into *Grettis Saga*: "Just over this gap is the site of the fabulous or doubtful Thorisdale of Grettis Saga: and certainly the sight of it threw a new light on the way in which the story-teller meant his tale to be told" (*Journal* 77).

And, again, writing of Hítardal: "It was such a savage dreadful place, that it gave quite a new turn in my mind to the whole story and transfigured Grettir into an awful and monstrous being, like one of the early giants of the world" (*Journal* 149).

The route that Morris and his party followed on the first expedition would seem to have been dictated by the sagas, chiefly *Njal's Saga*, *Laxdale Saga*, *Eyrbyggja Saga*, and *Grettis Saga*, but many of the minor ones also. The journey took him west from Reykjavík, then north to Vatnsdalr, turning east to circuit the Snaefell peninsula and continuing southeast to Thingvellir, the culmination of his journey and the place which, because of its many associations with sagas and history, seems to have given him the greatest pleasure. That the saga sites were his major concern is evident throughout, as references to them punctuate the text in rapid succession, indicating incidentally the breadth of Morris's reading of the sagas and the keenness of his memory in recalling them in detail. Much of the landscape is seen solely in terms of the sagas. Thus, typically: "We are come in Viðidal now, and behind us to the north-east can see the hills of Langdale, the main scene of the Bandamanna Saga: before us in a slope with a stead called Borg, the place of the Saga of Finnbogi the Strong" (*Journal* 94).

The singlemindedness of Morris's interest is made plain during the detour that the party made to view the geysers. Morris makes his displeasure felt as soon as they set up camp for what is to be three full days:

> Understand I was quite ready to break my neck in the quality of pilgrim to the holy places of Iceland: to be drowned in Markfleet, or squelched in climbing up Drangey seemed to come quite in the day's work; but to wake up boiled while one was acting the part of accomplice to Mangnall's Question was too disgusting. (*Journal* 67)

Though there follow interesting accounts of the eruptions of the two geysers Strokkr and Great Geyser, Morris regards the geysers as of no real interest except to geologists, and his descriptive account contrasts markedly with the technical explanations and descriptions of Lord Dufferin, who had visited the site some fifteen years earlier.[9] At the end of the detour, Morris makes no attempt to conceal his delight: "We were all in high spirits, I in special I think, for I had fretted at the delay in this place sacred principally to Mangnall, and there had seemed a probability of the expedition being spoiled or half spoiled" (*Journal* 74).

Morris's second journey (recorded in the *Diary*) is only partly geared to the sagas. The territory of *Njal's Saga* (west of Reykjavík) is briefly revisited, though by a different route, and the party then heads north through the glaciers to the north coast, via Lake Myvatn and Akureyri. Once among the glaciers, the land is uninhabited and free of saga associations. Morris's account of the journey here concentrates on the terrain and the changing (usually grim) vistas. The second trip gives the impression to a large extent of an endurance test. Morris is here traveling without the Icelander Magnússon and thus occupies a more prominent and responsible position in the party.

His relationship with the guides is closer than it had been in 1871 and we hear more of them in the *Diary* than in the *Journal*. In fact, he had met one of them, Jón Jónsson, on his previous visit and had scarcely remembered him, though the two men were later to become close friends. Morris, too, has by now gained in self-confidence and there is little of the nervousness that he had sometimes displayed in 1871. Some tension is built up as he looks forward to the awesome prospect of the journey north between the glaciers, via Sprengsandr ("the sands"), but "the sands" are quickly crossed and a sense of anticlimax is felt. Similar tension was present in the *Journal*, as a faint-hearted Morris wondered again and again whether or not to risk the precipitous Bulandshöfði on Snaefellsness. Morris overcame his fears but, once completed, there was again a sense of anticlimax:

> I for my part [was] well contented that the danger was little or nothing, if a little ashamed that my imagination had made much of it, C.J.F. rather disappointed, I think, and Evans scornful of the whole affair: by which you may see, I suppose, that I ought not to have spoken of it as a perilous pass at all. (*Journal* 134)

As is evident from the above, neither the *Journal* nor the *Diary* is without excitement, suspense, and humor. Morris frequently pokes fun at himself, at his pretensions, and at his shortcomings. He is often reluctant to rise at the early hour agreed, he describes difficulties with his Icelandic, he tells us how he was called "fat," he quarrels with Faulkner over Faulkner's snoring and he takes pride in his role as cook to the party. Although Morris does not portray himself in a particularly flattering light, the picture of him that emerges is of a good-humored and determined traveler, driven (on his first journey at least) by a singleminded , possibly selfish, desire to experience the Iceland of the sagas. The difficulties of travel in Iceland in the late nineteenth century emerge clearly enough from the accounts in both *Journal* and *Diary*. Roads are virtually nonexistent, and a team of ponies is indispensible, as are the services of local guides. At all times the travelers depended on the hospitality (never refused) of local farmers for accommodation or a place to pitch their tents. Morris was only concerned to report that which interested him most, and social information, where it occurs, is always incidental. Hay-making, sheep-farming, whaling, fishing, trade in wool and salt fish are noted when noticed, but none is investigated in any way. Poverty, wealth, and learning are remarked in the inhabitants whose paths briefly cross that of Morris, but few conclusions are drawn, except insofar as they relate to the sagas. Thus, in an extended and untypical bout of reflection in Laxdale, Morris laments the loss of those passions that fired the saga events, though they are still keenly remembered. Life in Iceland would seem to have changed, despite the unchanged landscape. "What littleness and helplessness has taken the place of the old passion and violence that had place here once," he writes, and, further, "whatever solace your life is to have here must come out of yourself and these old stories" (*Journal* 103).

Morris is here expressing the view of the disappointed Romantic confronted with reality. His mistake is to compare the fictionalized exaggerations of events and people of the distant past with a contemporary social reality that he, on the evidence of the *Journal*, seems at the most only superficially to have explored. Small wonder that he was disappointed—a disappointment with the modern Icelander which he repeated, in a slightly different spirit, in a long letter to Charles Eliot Norton. He writes almost affectionately of the poor, lazy Icelander, redeemed by his interest in the past:

> Then the people: lazy, dreamy, without enterprise or hope: awfully poor, and used to all kinds of privations—and with all that, gentle, kind, intensely curious, full of their old lore, living in their stirring past you would say, among dreams of "Furor Norsmanorum" and so contented and merry that one was quite ashamed of one's grumbling life—wasn't there something delightful & new about that also? (*Letters* 152)

Although both *Journal* and *Diary* have been praised for their clarity and the absence of archaism and artificiality, it must be admitted that, despite the descriptive highlights and passages of interesting anecdote, much of both the *Journal* and *Diary* is too plain, as Morris describes each stage of each journey, whether or not it was interesting, whether or not anything happened. Morris's zeal for thoroughness here seems to militate against total success. Large parts of his accounts can mean little, unless read alongside a map of the itinerary. (Wisely, one was included with the published edition.) The following is not untypical:

> Then passing by our old camp we follow up a willowy stream that runs under bents edging a sandy plain somewhat willow-grown also, with Skialdbreið ever on our left, looking no otherwise than we saw it weeks ago from the east side of it, for in short it is quite round. Then over a neck of shale and rock called Trollahals (Troll's Neck) into a great wide sandy valley, going utterly waste up to the feet of Skialdbreið, and now with a small stream running through it. We are now turning round Skialdbreið, and can see on his south-west flank two small hills lying that are perfect pyramids to look at from here. We are drawing near to the spurs of Armansfell now, and the wide plain narrows as a hill on our left shuts out the view of Skialdbreið, and then we are in a great round valley of dark

brown sand as flat as a table and almost without a pebble on it: the shoulder of Armansfell, the haunt of the land-spirits, rises on the south-west of the valley, and in the corner is a small tarn. (*Journal* 166–67)

Morris's visits to Iceland had served their purpose and he felt no need to return; the landscape had impressed him deeply and he retained strong memories of it for the rest of his life. "Do you know," he wrote to Aglaia Ionides Coronio, "I feel as if a definite space of my life has passed away now that I have seen Iceland for the last time. . . . surely I have gained a great deal and it was no idle whim that drew me there, but a true instinct for what I needed" (*Letters* 198). His interest in Iceland had deepened and its influence continued to shape his work and in 1876 he published his verse narrative *The Story of Sigurd the Volsung and the Fall of the Niblungs*, a recreation of the *Volsunga Saga*, which he had earlier translated with Eiríkr Magnússon. Despite his many other activities, Morris continued, with Magnússon, to translate Icelandic saga texts and between 1891–95 the five volumes of the Saga Library were published. These included *The Story of the Ere-Dwellers* (*Eyrbyggja Saga*), much of the territory of which he had visited in 1871, and, in three volumes, *The Stories of the Kings of Norway* (*Heimskringla*) by Snorri Sturluson, who is often referred to in the *Journal* and whose bath at Reykholt Morris had described in detail and bathed in.

The Icelandic influence in Morris's writing was more pervasive after his visits, and in his own verse abiding memories are revived in the poem "Iceland First Seen" (published in 1891 in *Poems by the Way*), in which he asks why one should wish to visit so unpromising a land. Inevitably, the answer lies in the sagas:

> Why do we long to wend forth
> through the length and breadth of a land,
> Dreadful with grinding of ice,
> and record of scarce hidden fire,
> But that there 'mid the grey grassy dales
> sore scarred by the ruining streams
> Lives the tale of the Northland of old
> and the undying glory of dreams?

The poem opens with a series of images of Iceland, as Morris first remembered seeing it from the sea, with its "toothed rocks" and mountains "all cloud-wreathed and snow-flecked and grey." Thereafter, Morris becomes more reflective: it is a land with a hidden past "as some cave by the sea where the treasures of old have been laid," a land, in fact, that awaits renewal, as "the spouse of a God" (no doubt Frigg, wife of Odin) is reunited with her son Balder. Here, in language that is relatively free from archaisms, Morris has used his memories of an apparently barren land to express his respect for the past and his hope for the revival of the spirit that animated it. In the same collection is "Gunnar's Howe above the House at Lithend," in which Morris reflects on the tomb of Gunnar (one of the heroes of *Njals Saga*), which he visited in 1871. Gunnar may not be remembered by many, Morris suggests, but the moral values expressed in his life and death ("the gladness undying that overcame wrong") are enduring and permanent. The words of Gunnar unite past and present and "bridge all the days that have been." In both poems, Morris is demonstrating and even justifying his belief that an awareness of the past is appropriate to the spiritual life of his own age.

However, it is in the later prose romances that Morris makes the most creative and imaginative use of his memories of Iceland. Though none is set in Iceland (or, in fact, in any identifiable land), there are frequent reminders of scenes described in both *Journal* and *Diary*. Several of these romances have as their structure a quest in which a young hero (generally) is obliged in the medieval manner to undergo a journey, which is part quest, part test, in order to establish himself as in some way special. In describing these journeys, Morris pays the same close attention to topographical detail that characterizes the *Journal* and *Diary*. The landscapes through which the hero travels are mostly a mixture of the idyllically pastoral and the ruggedly hostile and are often endowed with symbolic value, in that the hero must usually, at some stage in his journey, traverse dangerous wastelands and mountains as a part of his test. Morris's descriptions of these suggest that he is recalling his experiences of the Icelandic landscape, sometimes quite closely, at other times in a general way. In *The Well at The World's End* (1896), for instance, the hero and his companion have to cross a lava field:

> for betwixt them and the ridge . . . stretched a vaste plain, houseless and treeless . . . like a huge river or firth of the sea it seemed, and such indeed it had been once, to wit a flood of molten rock in the old days when the earth was a-burning. (*The Well at the World's End* bk. 3, ch. 8)

Morris recalls seeing several such lava fields in his *Journal* and *Diary* (as at Bolavellir, *Journal* 30, Thingvellir *Journal* 168, and Hliðarhæli *Diary* 229–30), any one of which could have been his inspiration here. However, in *The Well at the World's End* the travelers next encounter "the glaring of earth-fires" and "a very pillar of fire rising up from a ness of the mountain wall," which is unlike anything that Morris saw and is likely to be his idea of what a volcanic eruption would have looked like. Further parallels between Iceland as described in the *Journal* and *Diary* and the prose romances can be found in *The Story of the Glittering Plain* (1890), where Morris's description of the Isle of Ransom recalls his description in the *Journal* of the Faroes seen

92

from the north (*Journal* 17–18). In the mountains beyond the Glittering Plain, several passages recall Iceland, as Morris had earlier described it. Thus:

> he was high up amongst the mountain-peaks: before him and on either hand was but a world of fallow stone rising ridge upon ridge like the waves of the wildest of the winter sea. (*The Story of the Glittering Plain* ch. 17)

> but close under the aforesaid mountain-spur, a huge mass of black cliff, with a wild sea of lava tossing up into great spires and ridges landward of it, and at the back of that mountains and mountains again. (*Journal* 125)

In *The Roots of the Mountains* (1889), which is more epic than romance, mountain wastes are again featured, and here we have a clear indication in Morris's use of the name "Shield-broad" that he has an actual Icelandic location in mind:

> and beyond these western slopes could men see a low peak spreading down on all sides to the plain, till it was like to a bossed shield, and the name of it was Shield-broad. Dark grey was the valley everywhere, save that by the side of the water was a space of bright green-sward hedged about toward the mountain by a wall of rocks tossed up into wild shapes of spires and jagged points. (*The Roots of the Mountains* chap. 41)

> we see ahead and to our left the wide spreading cone of Skialdbreið (Broad-shield) which is in fact just like a round shield with a boss; running south from its foot is a rent and jagged line of hills. (*Journal* 76)

Other reminders of the Iceland of the *Journal* and the *Diary* are to be found in *The Sundering Flood* (1897), the last of Morris's romances, which has, according to May Morris, its origins in a modern Icelandic novel. Furthermore, she writes:

> the description of the sheer cliffs and the black water . . . take one back to the early days of (Morris's) Icelandic travel where the first sight of volcanic mountain heights seemed as much to overwhelm him with their terror as to move him by the majesty of their untrodden mysteries.[10]

In general, however, *The Sundering Flood* is set in the usual undefined land, a mixture of landscapes both harsh and pastoral, and at a time that is vaguely medieval. The language, too, is elaborate, full of archaisms, placing the

93

events in some distant past. The tale is one of young love triumphing over obstacles both geographical and human and at the heart of the story is the sundering flood, the river that divides them. In part, the setting recalls the Bairns' Force above Reykholt, which Morris visited in 1871:

> It is a wild place enough: a mile below Gilsbank White-water is about as wide as the Thames at Reading, two small rivers come into it between this and that, and here is all the rest of it shut up between straight walls of black rock nowhere more than twenty feet across; far up the gorge we can see the mountains towering up, and the white dome of the great Jokul beyond everything. (*Journal* 160–61)

> As for the Flood itself, it is now gathered into straighter compass, and is deep, and exceeding strong; high banks it hath on either side there of twenty foot and upward of black rock going down sheer to the water; and thus it is for a long way, save that the banks be higher and higher as the great valley of the river rises towards the northern mountains. (*The Sundering Flood* ch. 1)

The Cloven Knoll, where the two young lovers of *The Sundering Flood* first sighted each other, also bears some resemblance to the Bairns' Force:

> About the narrowest of it (the river), where it is certainly not ten feet across, the rocks stretch out to meet each other, overhanging the stream like the springings of some natural arch; which indeed the story of the place says was once complete, but that a certain witch once lured two children of her enemy to cross the place, and then raised a wild storm which swept away them and the keystone of the bridge: wherefore is the fall called the Bairns' Force. (*Journal* 161)

> Then came the straight passage of water, some fifty feet across, and then the bank of the eastern side, which, though it thrust not out, but rather was as it were driven back by the stream, yet it rose toward the water, though not so much as the ness against it. It was as if some one had cast down a knoll across the Sundering Flood, and the stream had washed away the sloped side thereof, and then had sheared its way through by the east side where the ground was softest. Forsooth so it seemed to the Dalesmen, for on either side they called it the Bight of the Cloven Knoll. (*The Sundering Flood* ch. 9)

The suggestion here is that Morris has retained a strong visual memory of the Icelandic landscape, which he had visited and described some twenty years earlier, and that he had adapted and recreated it to suit the purposes of his tales, assimilating to them other details, drawn both from his own imagination and from other landscapes (chiefly English) with which he was familiar. The result in the prose romances are landscapes that are convincingly real, though like no single place. These landscapes are thoroughly suited to Morris's romantic tales, tales that are at once familiar, universal, and tied to no particular place or clearly defined age.

Morris's experience of Iceland seems also to have played some part in his developing social awareness and his later political commitment, though there is little sign of this in either *Journal* or *Diary*. The sagas had given him innumerable illustrations of fiercely independent men fighting against overwhelming odds and his visit to Iceland had shown him, he writes in 1883, perhaps rather unexpectedly, "that the most grinding poverty is a trifling evil compared to the inequality of the classes."[11] His special sympathy for the Icelanders also drew him into active work on their behalf after a series of natural disasters in 1875 and 1882, work that was not entirely uncontroversial and that required him to develop both diplomatic and campaigning skills.[12] Iceland, it seems, needed Morris, just as much as he needed Iceland.

NOTES

Quotations from Morris's works are from the following editions: *Journals of Travels in Iceland*, vol. 8 of the *Collected Works of William Morris*, ed. May Morris with additional notes by Eiríkr Magnússon (London: 1911); *Poems by the Way* (London: 1891); *The Sundering Flood*, vol. 21 of the *Collected Works of William Morris*, ed. May Morris (London, 1913); *The Collected Letters of William Morris. 1848–1880*, vol. 1, ed. Norman Kelvin (Princeton, NJ: 1984).

1. Thomas Carlyle, "The Hero as Divinity. Odin. Paganism. Scandinavian Mythology," *On Heroes, Hero-Worship and the Heroic in History* (London: 1841), Lecture 1.
2. For a fuller account and discussion of Victorian writers and Iceland, see Karl Litzenberg, "The Victorians and the Vikings: A Bibliographical Essay on Anglo-Norse Literary Relations," *The University of Michigan Contributions in Modern Philology* (1947): 1–27.
3. Stefán Einarsson, "Eiríkr Magnússon and his Saga Translations," *Scandinavian Studies and Notes* 13 (1934): 24. Morris's working methods are also recorded in this account.
4. Peter Faulkner, ed., *William Morris. The Critical Heritage* (London: 1973), 160–61.
5. Of the tales published in *Three Northern Love Stories and Other Tales*, *The Story of Gunnlaug the Worm-Tongue* (*Gunnlaugs Saga*) was first published in the *Fortnightly Review* for January 1869 and the story of *Frithiof the Bold*

(Friðjófs Saga) was first published in *Dark Blue* for March–April 1871.

6. Quoted in Ann Thwaite, *Edmund Gosse: A Literary Landscape 1849–1928* (London: 1984), 110.

7. The volume comprises *A Journal of Travel in Iceland 1871* and *A Diary of Travel in Iceland 1873*.

8. Jan Morris, *Icelandic Journals by William Morris* (Fontwell, Sussex, Eng.: 1969), xix. This is a reprint, with a new introduction, of the *Collected Works* edition.

9. Lord Dufferin, *Letters from High Altitudes* (London: 1856), letter 7.

10. May Morris, *The Collected Works of William Morris*, vol. 21 (London: 1913), xi.

11. Quoted in John Purkis, *The Icelandic Jaunt: A Study of the Expeditions made by Morris to Iceland in 1871 and 1873* (Kew, Eng.: 1962), 28.

12. For a discussion of Morris's relief work, see Richard L. Harris, "William Morris, Eiríkr Magnússon, and Iceland: A Survey of Correspondence," *Victorian Poetry* 13 (1975): 119–30. Also the same author's "William Morris, Eiríkr Magnússon, and the Icelandic Famine Relief Efforts of 1882," *Saga Book of the Viking Society for Northern Research* 20 (1983): 31–41.

HOPKINS IN DUBLIN

Warren Leamon, University of Georgia

Sometimes irony develops slowly. In the late 1820s the suppression of ten Irish bishoprics by Parliament seemed to Keble, Froude, and Newman a sure indication of disestablishment. Their response, a spirited effort to revive and maintain what they saw as the old Anglo-Catholic tradition, led to the Tracts and the controversy that they provoked. In a sense, then, Ireland was the cause of the Oxford Movement—the proximate cause, at any rate. And Newman would never have gone down to defeat in Ireland in his attempt to establish a university if he had not become, as one result of his involvement in the Oxford Movement, a Roman Catholic. And if there had been no remnant of the Oxford Movement existing at Oxford when Gerard Manley Hopkins entered in 1861, he quite likely would never have become a Roman Catholic, hence never a Jesuit priest, hence never a professor of classics in the ruins of Newman's university in Dublin.

This is merely a small example of how Ireland runs through English history like Fate in a Greek play, producing one catastrophe after another. Surely one of the starkest symbols of the tragic relationship between the two countries is the grave of Hopkins in Glasnevin Cemetery in Dublin: the quintessential, even chauvinistic, English poet, dead in a city he hated perhaps more than he hated other cities, in a country, probably the most Roman Catholic country on earth, that he, a Roman Catholic priest, never really understood, and laid out with Parnell and O'Connell and the martyrs of '16 and '48 and '98.

Whatever happened to Hopkins during his five years in Ireland must be examined in the context not only of his life but in the context of Victorian society as well. Thus, though he arrived in Dublin when the Land War was raging, boycotts and tenant strikes were tearing the countryside apart, and Parnell, attempting to move Parliament toward a Home Rule bill, was raising the threat of armed revolt, Hopkins, in his first letter to Robert

Bridges, does not bemoan politics or the Irish; rather he notes that "The house we are in, the College, is a sort of ruin . . . Dublin itself is joyless place and I think in my heart as smoky as London is: I had fancied it quite different" (*The Letters of Gerard Manley Hopkins to Robert Bridges* [henceforth, *Letters*] 190).

Hopkins was familiar with industrial cities, particularly Liverpool, so he probably "had fancied" Dublin a pre-industrial city—an English country town on a large scale, the first of many illusions that were to be shattered. Like most other things in Hopkins's life, we can only guess at what those illusions about Ireland were, but in all likelihood they were the typical English ones: that Ireland was a country of emotional nationalist leaders who were misguiding basically loyal farmers; a picturesque primitive landscape dominated by the great houses of Anglo-Irish aristocrats who, like the peasants they ruled, had weaknesses but were on the whole faithful to the principles of Empire. The main difference between Hopkins and most other English intellectuals and writers who had gone to Ireland before him was that he was a Roman Catholic. At first this allegiance may have produced other illusions, in particular, the comforting thought that he was going to a Catholic country. But it wasn't long before he realized that the Church was deeply involved in the turmoil, from grassroots parish priests caught between compromise and oppression to the nationalist bishops of Dublin and Cashel.

Yet his first response had nothing to do with politics or religion; it had to do with how "smoky" Dublin was, with how "inconveniently far off" the Phoenix Park was, and with the barely comforting observation that "there are a few fine buildings" (*Letters* 190). We tend to think of the "Terrible Sonnets" as embodying his reaction to his "exile" and forget that while in Dublin he also wrote "Tom's Garland" and "Harry Plowman."

We shouldn't be surprised that the physical aspect of Dublin was the first thing about Ireland that depressed Hopkins. Despair over what "progress" was doing to nature runs though his poetry from beginning to end. Doubtless this despair resulted in large measure from his aestheticism; his notebooks, journals, sermons, letters, and retreat notes reveal a temperament acutely, even neurotically, sensitive to the most subtle nuances of the environment. But Hopkins does seem to consider man's corruption of his environment more a spiritual than a social transgression. And just as the Incarnation represented for Hopkins the heart of Christian faith, so the Industrial Revolution was the heart of the assault on God, out of which flowed the blood that soaked the English countryside. When, from the perspective of his later life, we examine the flurry of letters generated by his conversion, one seemingly enigmatic sentence stands out. Pusey, to explain why he refused to see Hopkins, appended the following sentence: "I agree

98

that the poor ought to be able to discern the church; but I think the poor feel what we feel as difficulties as much as we" (*Further Letters of Gerard Manley Hopkins* [henceforth, *Further*] 400). Of course, I don't believe that Hopkins's social consciousness was responsible for his conversion. But the opposite interpretation, that his conversion contradicted his social views, is just as mistaken. Hopkins sensed a relationship between the politico-economic system and the spiritual well-being of the country; against the overwhelming pressure of his class and his own patriotism, which bordered on chauvinism, he opted for Rome.

This social consciousness shows up early in his decision to become a Jesuit. Why Hopkins joined the Society of Jesus has been much debated. The most popular view is that—whether consciously or unconsciously—he was drawn to the Society's rigid discipline to restrain his sensuality and the feelings of guilt that resulted from it. It is just as likely, however, that Hopkins was drawn to the Society for two other reasons: it was the most "English" of the religious orders in England (that is, the order most favored by the "Old English Catholics"), and it was deeply involved in evangelical work. The order's numbers grew rapidly during the last half of the century; it was becoming heavily involved in education, and it was making inroads into the working classes of the North, where there was a solid base of Irish immigrants.

Early in his career as a Jesuit, in a letter lecturing Bridges on the plight of the poor, Hopkins asserts that the industrial age has produced a society, an "iniquitous order," held tenuously together by the "secular statesman." "Horrible to say, in a manner I am a Communist," he writes, and goes on to excoriate England as a country in which the lower classes produce wealth for the upper classes but do not share in it themselves. Being uneducated, the poor cannot appreciate the older traditions of the country and thus cannot be blamed for wanting to destroy the system that grinds them down. "The more I look, the more black and deservedly black the future looks," he concludes (*Letters* 27–28).

What is striking about the letter is that Hopkins couches his argument in purely secular terms and though he professes to reject the "methods" of the Communists, he says of the Communist ideal, "Besides, it is just." But while he flirts with the idea, so common today, that only revolution can transform society, he never takes the final step of using Christianity to justify social upheaval, and six years later in "The Wreck of the Deutschland" he expounds the incarnational approach to social problems that will characterize his thinking for the rest of his life.

In "The Wreck of the *Deutschland*" Hopkins achieves a fine balance between inner and outer reality, between self and nature. And during the years between "The Wreck" and his going to Ireland, this balance made it possible for him to hold fast to his love of England even as he lamented its spiritual failure. His bouts with depression are generally reflected in his poetry through his deep conviction that the ills of modern man result from the perversion of the relationship between man and nature. A mining town in Wales reveals the results of embracing Darwin's view of man; the industrial age had ruined Oxford by destroying the subtle balance between the man-made and the natural; the destruction of trees is lamented and "the weeds and the wilderness" are cheered; the mystery of the resurrection of the flesh transforms the caged bird and the grim work of the blacksmith; and again and again the labor and things of man achieve beauty when they achieve harmony with the natural world. Whatever alienation he experienced from the country he loved resulted from the fact that England had rejected him, not that he had rejected England. But as Christ stood at the center of his society—the Society of Jesus—in the Eucharist, so the Holy Ghost brooded over England, able to transform it at any moment. The aesthetic version of medievalism embodied in the pre-Raphaelites and in Ruskin's theories translated easily into his religious medievalism, his dream of that Catholic England that had produced Duns Scotus.

But his secular ideal remained the Victorian gentleman and what sustained him during these years was his strong sense of duty. His failures within the order—failures in his studies, his teaching, his preaching—were made bearable by his faith in God's ability to transform his "sheer plod." This faith was reenforced by the Society of Jesus, which tempered examination of the soul with a strong devotion to community. Moreover, Hopkins perceived no *fundamental* clash between the spiritual values of the Catholic Church and the moral values of his country. After all, his sense of duty derived as much from his Victorian background as it did from his love of Christ and the Church. Put in medieval Catholic terms, England possessed the moral virtues that made it, in Hopkins' view, a great force for civilization in the world. His Church was not opposed to those virtues; on the contrary, it sought to elevate them by turning the country back to the theological virtues it had abandoned. Hopkins's emotional instability, his excessive scrupulosity, his tendency to depression were held in check by the delicate balance he perceived among self, Church, and country.

In Ireland the balance fell apart. When Hopkins disembarked in Dublin on a rainy, windy day in February 1884, he entered a country whose social and political situation was as uncertain as its weather. And it was the weather along with the physical aspects of the city that first affected him, but as we have seen, his extreme sensitivity to his environment was bound up with his spiritual and sociopolitical obsessions, and it was not long before complaints about the weather and the unpleasantness of the city give way to highly ambiguous complaints about his health. In a sense, the fear of the heavy burden of examinations wears him down even before he begins grading the papers.

In fact, he spent much of his time after his arrival not in the confines of Dublin but in the Irish countryside. In April he writes to Bridges from Clongowes Wood, then in July from Galway and remarks, "I have been through Connemara, the fine scenery of which is less known than it should be . . . Furbough House stands amidst beautiful woods, an Eden in a wilderness of rocks and treeless waste. The whole neighbourhood is most singular" (*Letters* 193).

In September, back in Dublin, he sends Bridges a postcard: "I am in the very thick of examination work and in danger of permanently injuring my eyes" (*Letters* 198). In October he writes to Dixon that he is "drowned in the last and worst of five examinations. I have 557 papers on hand: let those who have been thro' the like say what that means" (*The Correspondence of Gerard Manley Hopkins and Richard Watson Dixon* [henceforth *Dixon*] 123). In November, at the end of a long lively entertaining letter concerning "married life," poetry, and music he apologizes for his dullness, saying that he is "wearifully tired" (*Letters* 200). Throughout the early part of 1885 (some of which time he spends in the country) his letters are lively, full of literary criticism and observations on art and nature in general. A letter to his mother indicates that he was by no means uninterested in the political situation. After a light and humorous description of a political rally in the Phoenix Park that he and another priest went to ("I fancy it was rather compromising"), he falls prey, as Englishmen are wont to do, to the old cliché of the stage Irishman: "Excitable as the Irish are they are far less so than from some things you would think and ever so much froths off in words." Then, characteristically, his tone changes:

> Though this particular matter did not disturb me, yet the grief of mind I go through over politics, over what I read and hear and see in Ireland about Ireland and about England, is such that I can neither express it not bear to speak of it. (*Further* 170)

On St. Patrick's day he writes in his spiritual diary, "Ask his help for Ireland in all its needs and for yourself in your position" (*Sermons* 260). And at the end of March 1885, a year after his arrival in Ireland, his political and social concerns break momentarily through his complaints about his work and his jealous pleas for attention from Bridges. In a long letter containing detailed criticisms of Bridges's *Nero* he suddenly remarks in an aside: "But there is no depth of stupidity and gape a race could not fall to on the stage that in real life gapes on while Gladstone negotiates his surrenders of the empire" (*Letters* 210). In his next letter he writes of "that coffin of weakness and dejection in which I live, without even the hope of change" (*Letters* 215), and a little over a month later he writes, "I think that my fits of sadness, though they do not affect my judgment, resemble madness. Change is the only relief, and that I can seldom get" (*Letters* 216).

Between these last two letters, on April 24, he writes to his old college friend Mowbray Baillie. This letter is crucial in the glimpse it gives us into the complexity of Hopkins's discontent in Ireland. The opening sentence is striking for its unsuccessful attempt to temper morbidity with wit: "I will this evening begin writing to you and God grant it may not be with this as it was with the last letter I wrote to an Oxford friend, that the should-be receiver was dead before it was ended" (*Further* 254).

But as Hopkins goes on to write about Geldart and Nash, two Oxford friends he believes committed suicide, his tone changes: "Three of my intimate friends at Oxford have thus drowned themselves, a good many more of my acquaintances and contemporaries have died by their own hands in other ways: it must be . . . a dreadful feature of our days" (254). He returns to the letter a week later and, after some entertaining recollections, suddenly writes:

> I think this is from a literary point of view (not from a moral) the worst letter I ever wrote to you. . . . You will wonder I have been so long over it. This is part of my disease, so to call it. The melancholy I have all my life been subject to has become of late years not indeed more intense in its fits but more distributed, constant, and crippling. One, the lightest but a very inconvenient form of it, is daily anxiety about work to be done, which makes me break off or never finish all that lies outside that work. It is useless to write more on this: when I am at my worst, though my judgment is never affected, my state is much like madness. (256)

Then he turns immediately to

> the verses I shewed you. . . . Those verses were afterward burnt and I wrote no more for seven years; then, it being suggested to write something I did so and have at intervals since, but the intervals are now long ones and the whole amount produced is small. And I make no attempt to publish. (256–57)

One can discern a connection between the "disease" and the fact that "the whole amount produced is small"; and one can make the connection with the letter to Bridges written a little later in which he says, "I have after long silence written two sonnets . . . if ever anything was written in blood one of these was" (*Letters* 219). But Hopkins twists once again:

You said, and it was profoundly true then, that Mr. Gladstone ought to be beheaded on Tower Hill and buried in Westminster Abbey. Ought he now to be buried in Westminster Abbey? As I am accustomed to speak too strongly of him I will not further commit myself in writing. (*Further* 257)

This final observation may account for the conclusion: "Much could be said about Ireland and my work and all, but it would be tedious." One can, I think, assume that Hopkins included Ireland in those "surrenders of the empire" that Gladstone should be beheaded for.

So when, at the end of the summer in one of his most famous letters (*Letters* 220–22) Hopkins remarks to Bridges that he will soon send him some sonnets that "came like inspirations unbidden and against my will," we know that there is no one source for whatever despair informs those poems—if, indeed, any despair informs them. After all, the letter in which he refers to them is one of the most ambiguous he ever wrote—a kind of metaphor for his Dublin years, not to mention his life. And at the end of it he insists, "I do not despair."

He begins this letter by saying that he has "just returned from an absurd adventure" that he fears might be "compromising" during which a "hare-brained fellow" detained him overnight on a yacht, but which "was fun while it lasted." He is in a good mood apparently because he has been in England visiting his family in Hampstead and Sussex "in a lovely land-scape," and Coventry Patmore in Hastings. "I managed to see several old friends and to make new ones." He objects to Bridges's "contemptuous opinion" of Barnes's poems, insisting that Barnes "is a perfect artist and of a most spontaneous inspiration." He then turns to "an old question of yours," whether he (Hopkins) is "thinking of writing on metre." He hopes to be able to do so and the contemplation of the possibility leads him to remember his fits of anxiety:

> if I could get on, if I could but produce work I should not mind its being buried, silenced, and going no further; but it kills me to be time's eunuch and never to beget . . . soon I am afraid I shall be ground down to a state like this last spring's and summer's, when my spirits were so crushed that madness seemed to be making approaches. (222)

It's difficult to know how seriously to take this lament since he was saying practically the same thing before he ever began any of his work in Ireland. In this letter, in fact, sandwiched between a lighthearted defense of Barnes's poetry and some musing on the possibility of "writing on metre"—and just after he has said that he will soon send some sonnets—he interjects: "And in the life that I lead now, which is one of a continually jaded and harassed mind, if in any leisure I try to do anything I make no way—nor with my work, alas! but so it must be" (221). In November 1885, in a letter to his mother (*Further* 173), we discover that ordinary academic politics have something to with his complaints.

Thus the Terrible Sonnets, along with "Spelt from Sybil's Leaves," may or may not be the product of despair; if they are then Hopkins's lifelong fits of depression (produced perhaps by his hypochondria, by guilt over sexual ambivalence and spiritual failure, or perhaps by a combination of the three) were exacerbated in Ireland by a combination of poor health, bad weather, hard work, politics, and academic in-fighting.

All the commentators on the Terrible Sonnets that I know of take the essentially Romantic position that the poems reflect an actual spiritual crisis and that the order of composition, though we probably will never know what it was, is somehow related to the development of the crisis. As I have tried to show, this may or may not be true. "Spelt from Sybil's Leaves," written in the fall of 1884, months before the Terrible Sonnets, is bleaker than the bleakest of the later poems while "To What Serves Mortal Beauty," written probably before but certainly no later than four of the Terrible Sonnets, reveals a coming to terms with the emotional and spiritual tempta- tions of this world as peaceful and complete as any "resolution" found in "My own heart let me more have pity on." And "Ashboughs," written at about the same time, is one of Hopkins's most beautiful celebrations of nature, more moving, perhaps, than some of the better known nature poems.

It seems most likely that in Ireland, Hopkins did not experience a spiritual crisis, a "dark night of the soul" (no matter how loosely we interpret that term)—rather something just the opposite happened: Ireland, because it made Hopkins an alien, provided him with the detachment he needed to gain insight into himself and England. What makes this difficult to perceive is that it took Hopkins himself a long time to perceive it (if, indeed, he ever did). Had he gone to India, South Africa, or America he would have known the initial shock of alienation for what it was; but like most Englishmen he conceived of Ireland not as a foreign country but as a primitive province. Even after he had been in Ireland for three years, he wrote to Newman, apparently complaining about the political situation, only to receive the following reply:

There is one consideration however which you omit. The Irish Patriots hold that they have never yielded themselves to the sway of England and therefore have never been under her laws, and have never been rebels. . . .

"An Eviction near Glenbeigh in Ireland" (1887), A. Forestier (*Illustrated London News* 90.111).

If I were an Irishman, I should be (in heart) a rebel (*Further* 413–14).

What Newman wrote was simple and self-evident, but Hopkins could never bring himself to accept the Irish position, though he did seem finally to understand that the Irish *had* a position. Thus he moves from naive attacks on the archbishops of Dublin and Cashel, accusing them of "robbery" and "rebellion" (*Letters* 251–52), to a kind of resignation: "What is to be done? Only one thing now: give them Home Rule. It will not end all our troubles, but at any rate they will be much worse without it" (*Further*, 181).

For many critics "To seem the stranger" (*Poems* 101) is central to an insistence upon a spiritual crisis. Thus Mariani (212–18)[1] considers it the beginning of a descent into darkness; Robinson (144–45)[2] finds it at the heart of desolation; Harris (8–10; 113–25)[3] finds in it evidence that Hopkins never rose out of his despair. The trouble is, the poem, the most "original" of the Terrible Sonnets—the one which, more than any of the others, makes the Terrible Sonnets unique, different from other works reflecting a "dark night of the soul"—really doesn't fit any scheme of development. If we abandon a search for an order of composition related to an actual concentrated experience of spiritual aridity, the poem reveals the complexity of Hopkins's Dublin years.

In the opening lines, the repetition of the word *stranger* is countered by "To seem," which suggests that his alienation is in some way apparent, not actual. The enigmatic phrase is also ambiguous: to *whom* does he seem the stranger? This ambiguity is intensified by the shifting meaning of "stranger," which in the first instance refers to Hopkins, but in the second instance to . . . whom? If we know the background of the sonnet, we understand that Hopkins is a stranger to the Irish (even after a year and a half) and the Irish are strangers to him. But the sudden shift to "Father and mother dear, / Brothers and sisters . . ." indicates that the strangers are his family. Thus his choice of Christ ("My peace/my parting") has made him a stranger to his family and they to him. The "my life" of the first stanza, then, refers not to his present existence in Ireland but to his whole life since his conversion.

At the beginning of the poem Hopkins is preparing to use his physical existence in Ireland as a metaphor for his Roman Catholic existence in life. "My lot" refers not only to "choosing lots" but to his lot in the sense of place, location (he uses a similar "real estate" metaphor in "Duns Scotus's Oxford"). The stroke in "My peace/my parting" is explained by Gardner as a method to "fetch out the painful paradox of *peace* and *parting*" (*Poems* 288). It seems to me just as likely that he uses it to distinguish the phrase from the rest of the line, "sword and strife"; that is, to make clear that "sword and strife" go with "parting" and that peace and parting and sword and strife are not balanced pairs. Both, in turn, are distinguished from "my lot, my life" in which the comma is used to create an ambiguity.

In the second stanza he turns abruptly to "England." The shift is not quite so abrupt as it seems, however, since we soon learn that England is his family, too, his wife, "whose honour O all my heart woos." His choice of Christ has separated him from his wife, just as it separated him from his mother and father and siblings, thus fulfilling Christ's command (Matt. 19:29). But at this point Hopkins refers to England as "wife to my creating thought." Why *creating thought*? Although "thought" is very broad and vague, it is hard to see how it could refer to either his academic or religious work in Ireland. Of course, knowledgeable readers connect it with the great poems of the late 1870s and early 1880s—poems usually inspired by the beauty of the English and Welsh countryside; and the letters of his Dublin years sometimes refer to his inability to write. But surely we are not to understand that Hopkins is in despair over writer's block.

This stanza also introduces a sexual metaphor that is to become more explicit as the years go by in Ireland: he is divorced from his wife and cannot conceive. Hopkins also suggests that his words ("pleading") would not find their mark, would not be heard by England, an idea picked up in the final tercet and expressed fully in the "dead letter" metaphor of "I wake and feel the fell of dark." The octet ends about where Milton's ends in "When I consider how my light is spent," with Hopkins lamenting his being "idle . . . where wars are rife." And, interestingly enough, we are over halfway through the poem before we encounter what could be considered a specific reference to Ireland.

"[W]here wars are rife" appropriately introduces Ireland: "I am in Ireland now." And just as appropriately, perhaps, the poem begins to fall apart. For now he says, "I can / Kind love both give and get," which seems to abandon the despair over his lack of "creating thought" and return to despair over separation from family. But then he seems to return to his inability to create with the reference to "what word / Wisest my heart breeds," which perhaps, also carries on the sexual metaphor; however, it could be that "word" is used here in the theological sense. At any rate the poem returns to the dead letter metaphor ("hoard unheard, / Heard unheeded") and ends with the highly unsatisfactory "lonely began," which is more of a gimmick than an eccentricity.

Harris (7), after a study of the manuscripts and texts of the Terrible Sonnets, concludes that "To Seem the Stranger" may have been the poem, of the four that "came like inspirations unbidden and against my will," that Hopkins began and ended with. From this Harris concludes that the sonnets follow no progression from descent to ascent. A less Romantic interpretation is that Hopkins kept fiddling with it because he could never bring the poem into focus. While the confusion of "No worst, there is none," "I wake and feel the fell of dark" and "Carrion Comfort" is a genuine spiritual confusion that paradoxically gives the poems their coherence, the confusion of "To seem the stranger" is existential. He first despairs over separation from family and country, then over an inability to create and finally over an inability to accomplish anything in Dublin. None of these sources of anguish is essentially new; what is crucial is the way the poem juxtaposes them.

The poem contains abandoned extended metaphors and an extended metaphor so subtle as to be ineffective. The family metaphor gives way to the failed writer metaphor which in turn gives way to the failed priest metaphor. The sexual metaphor is similarly inconclusive. The best case one can make for structure is the "location" metaphor, which begins obscurely with the play on the word "lot," continues through the juxtaposition of England and Ireland, and concludes with the suggestion that the speaker is still at the beginning of his journey. "[L]onely began," while it is aesthetically a failure, is a wonderful summing up of all the aesthetic failures of the poem.

Whatever the position of "To seem the stranger" in the order of the Terrible Sonnets, its treatment of family and politics certainly violates the usual vision of the "dark night of the soul" and gives the sonnet a secular concreteness such poems normally don't have. Dublin forced Hopkins to step back, to put his life into some kind of perspective, and the first result was brilliant incoherence.

The poems that Hopkins wrote for the next three and a half years in Dublin, up to his death from typhoid fever in May of 1889, show a slow development out of eccentricity and obscurity ("Tom's Garland," "Harry Ploughman," "Heraclitian Fire,") to a fine simplicity and directness ("St. Alphonsus Rodriguez," "Thou art indeed just," "The Shepherd's Brow," "To R.B."). Paralleling the development of his poetry was his coming to grips with Ireland and the political and social situation in the country. In February 1887 (*Letters* 251–52) he says of archbishops Walsh and Croke, "One . . . backs robbery, the other rebellion" but in resignation goes on to say:

> the people in good faith believe and will follow them. You will see, it is the beginning of the end: Home Rule or separation is near. Let them come: anything is better than the attempt to rule a people who own no principle of civil obedience at all. (252)

Confusion is apparent: how can people who "in good faith believe . . . and follow" be "a people who own no principle of civil obedience"? He would "be glad to see Ireland happy" but he believes that they cannot be and that the rebellion "has throughout been promoted by crime." He predicts a repeat of the Dublin Parliament which led to the '98 Rising, interpreting both movements as class struggles. "The ship I am sailing in may perhaps go down in the approaching gale: if so I shall probably be cast up on the English shore."

The old English and Empire prejudices are still there, embodied in paternalism, a paternalism that is even more apparent in a letter to Baillie written three days later (*Further* 281–83), in which he says, "Home Rule of itself is a blow for England and will do no good to Ireland. But it is better than worse things." He acknowledges that the "Irish had and have deep wrongs to complain of," but he professes to find it "strange" that reforms have not satisfied them even though he immediately observes that "the object of their undying desire and now of their flaming passion . . . is what they call Nationhood." In Hopkins's view independence will ultimately bring the Irish grief since, once again, they "own no principle of civil allegiance." He asserts that the position of tenant farmers is "more favorable than that of any other tenant farmers," yet still they want more. If they won't obey England or abide by the decisions of the Crown courts, then they won't obey any government. Like little children, "they must have Home Rule with all that it may cost both them and us . . . Gladstone is a traitor. But still they must have Home Rule."

By July 1887 he is more insistent on England's culpability: "It has always been the fault of the mass of Englishman [*sic*] to know and care nothing about Ireland, to let be what was there (which, as it happened, was persecution, avarice and oppression)." He believes as strongly as ever that "Home Rule is likely to come" and feels that it may "be a measure of a sort of equity and . . . a kind of prudence" (*Letters* 256–57).

The most one can say, then, is that Hopkins gained some insight into the complexity of the Irish problem and finally became resigned to Home Rule as the least of three evils (the other two being complete separation and military intervention on the part of England); but he always viewed the granting of Home Rule as evidence of the failure of English will, a failure that was leading to the collapse of Empire, and he found in Gladstone ("the Grand Old Mischief Maker") the embodiment of English folly and failure. And though he acknowledged that England had mistreated Ireland, he clung to the notion that such mistreatment resulted in part at least from the character of the Irish people, who, he believed, could only be ruled by a tyrant. At the end, shortly before his death, he was insisting, in the face of overwhelming evidence to the contrary, that the Pigot letters—the heart of an attack upon Parnell as an assassin—were genuine (*Letters* 256).

But Hopkins, despite his entanglement in the Irish mess, was not a political person and the most important effects of the Irish turmoil can be found in his growing awareness that some problems cannot be "solved." He learned through experience the truth of his assertion in the Terrible Sonnets, that patience is a "hard thing." For it is patience—the virtue of endurance —that underlies his last poems and gives us an idea of what might have been had he lived longer.

The final four poems—"St. Alfonsus Rodriguez," "Thou are indeed just," "The shepherd's brow" and "To R.B." (*Poems* 106-8)—constitute another version of the Terrible Sonnets. The sense of having been abandoned, the limitations imposed by the flesh, the need to be patient and endure—the central concerns of the earlier poems are the concerns of these. But these poems are more detached, more objectively controlled. The mountains of the mind, those "cliffs of fall / Frightful, sheer, no-man-fathomed" are now imaged in a lay brother who endured "the war within"; the disappearance of God, subject of self-torture in "I wake and feel the fell of dark," now produces a knight's chivalric plea to his lord; and physical self-disgust, an underlying reality of the Terrible Sonnets, now moves out into a general disgust in which the flesh signals man's fallen state and need for external help ("tame / My tempests there, my fire and fever fussy"). The sexual metaphors become more direct and more functional, culminating in the splendid extended metaphor that organizes "To R.B."

That Hopkins had achieved a new balance, more mature than the one that sustained him before he came to Ireland, is best seen in the juxtaposition of "Thou art indeed just, Lord" against "The shepherd's brow." Bridges thought the latter poem an aberration tossed off in a moment, but Gardner points out that Hopkins revised the poem several times (*Poems* 296). Thus his dwelling on what he considered the disgusting and squalid realities of the flesh does not drive him to despair and prevent him, in "Thou art indeed just, Lord," from treating the more serious sins of the intellect and spirit in a finely crafted sexual metaphor that concludes in the moving cry of the last line, "Oh Lord, send my roots rain." And in the last poem, "To R.B.," his lifelong devotion to the Incarnation as the central overwhelming mystery of Christianity produces an "explanation" of his inability to write poetry. After a transposition of Pentecost to the moment of inspiration that begins the act of creation, he then relates both to procreation, which issues in a stunning association of fire with the sexual act: "Sweet fire the sire of muse, my soul needs this; / I want the one rapture of an inspiration." One has to go back to the English poets of the seventeenth century—to Donne, Marvell, Crashaw—to find such an unforced mixing of the sexual and spiritual. Ireland, it seems, has taught him to be patient, to endure, to wait; the frantic ending of "To seem the stranger" now becomes "my winter world" and "with some sighs" he awaits spring.

110

But we, who know that Hopkins died scarcely a month after writing "To R.B.," receive the full force of "my winter world." His winter began with the gathering darkness of "Spelt from Sybil's Leaves" and extended through his years in Dublin—five years that are echoed in "To R.B.":

> Nine months she then, nay years, nine years she long
> Within her wears, bears, cares and combs the same:
> The widow of an insight lost she lives, with aim
> Now known and hand at work now never wrong.

NOTES

Quotations from Hopkins's works and letters are from the following editions: *The Correspondence of Gerard Manley Hopkins and Richard Watson Dixon* (London: 1955); *Further Letters of Gerard Manley Hopkins* (London: 1956); *The Letters of Gerard Manley Hopkins to Robert Bridges* (London: 1955); *Poems of Gerard Manley Hopkins* (4th edition, London: 1967); *The Sermons and Devotional Writings of Gerard Manley Hopkins* (London: 1959).

1. Paul L. Mariani, *A Commentary on the Complete Poems of Gerard Manley Hopkins* (Ithaca, NY: 1970).
2. John Robinson, *In Extremity* (Cambridge, Eng.: 1978).
3. Daniel A. Harris, *Inspirations Unbidden* (Berkeley and Los Angeles: 1982).

"CIVILIZATION" IN AFRICA, 1877–1900

Simon Gatrell, University of Georgia

In the year 1877 there occurred in Pretoria, the capital of the Transvaal in South Africa, a meeting between one of the grand old men of Victorian letters and a young colonial administrator who would become one of the best-selling novelists of the last quarter of the nineteenth century. The colonial administrator was Rider Haggard, and he had this to say about the encounter in his autobiography *The Days of My Life* (1926):

> Another noted man who visited us was Mr Anthony Trollope, who rushed through South Africa in a post-cart, and, as a result, published his impressions of that country. My first introduction to him was amusing. I had been sent away on some mission . . . and returned to Government House late one night. On going into the room where I was then sleeping I began to search for matches, and was surprised to hear a gruff voice, proceeding from my bed, asking who the deuce I was. I gave my name and asked who the deuce the speaker might be.
> "Anthony Trollope," replied the gruff voice, "Anthony Trollope."
> Mr Trollope was a man who concealed a kind heart under a somewhat rough manner, such as does not add to the comfort of colonial travelling. (1.136–37)

This meeting is symbolically important, because it came at a critical moment in the development of the British Empire, when the established laissez-faire policies that Trollope took for granted were giving way to the interventionist New Imperialism of which Haggard and his leader Theophilus Shepstone were embryonic representatives. In their writings about South Africa, Trollope and Haggard express fully the differences between old and new, but here it seems feasible to consider, as an indicator of these differences, their respective uses of the term "civilization."

113

Nowadays we are accustomed to using civilization as a relative term with some modifier (for example, Mayan civilization, industrial civilization), but for the Victorians, and perhaps for them alone, the word represented a single thing, approximating to the condition of developed Western European society and its offshoots across the oceans.[1] Any social state that did not conform to this model was more or less barbaric or savage. Trollope addresses the topic in a characteristic sentence from his book *South Africa* (1878), a sentence that introduces his discussion of the actions of Langalibalele, a black African ruler:

> When a Savage,—the only word I know by which to speak of such a man as a Zulu Chief so that my reader shall understand me; but in using it of Langalibalele I do not wish to ascribe to him any special savage qualities;—when a Savage has become subject to British rule and will not obey the authority which he understands,—it is necessary to reduce him to obedience at almost any cost. (1.333–34)

Trollope here makes a distinction between savage meaning non-civilized in general, and savage as denoting acts of especially brutal behavior in particular. When he comes to write about the Boers it is again the general sense of savage that he invokes:

> The Boer has become solitary, self-dependent, some would say half savage in his habits. The self-dependent man is almost as injurious to the world at large as the idle man. The good and useful citizen is he who works for the comfort of others and requires the work of others for his own comfort. (2.109)

Trollope would have been among those who considered the Boers "half savage." It is his understanding that work and the interchange of the products of work are the essential elements in civilization. He does not believe that it is necessarily the responsibility of the colonists to civilize either Boer or black, but if civilizing is to be done, there is only one way to do it—and that isn't through missionary endeavor:

> A little garden, a wretched hut, and a great many hymns do not seem to me to bring [a black] man any nearer to civilization. Work alone will civilize him, and his incentive to work should be, and is, the desire to procure those good things which he sees to be in the enjoyment of white men around him. (1.8)

If a man can be taught to want, really to desire and to covet the good things of the world, then he will work for them and by working he will be civilized. If, when they are presented to his notice, he still despises them,—if when clothes and houses and regular meals and education come in his way, he will still go naked, and sleep beneath the sky, and eat grass . . . and remain in his ignorance though the schoolmaster be abroad, then he will be a Savage to the end of the chapter. (1.323)

Trollope's emphasis upon work and the material benefits of a capitalist consumer economy to establish Empire and civilization was essentially shared by a central government, committed to free trade, which felt that colonies should be self-supporting, economically and militarily, and saw no special moral or political imperatives in maintaining them at the British taxpayers' expense if they could not maintain themselves. But both Trollope, and the secure, unquestioned view of civilization that he expressed, were fast becoming old-fashioned. When he encountered Haggard at Pretoria in 1877 he came face to face, not for the first time in his South African travels, with the idealism of the new movement.

Haggard's understanding of the concept civilization was more complicated and hardly involved at all the notion of work. In the same year, 1877, Haggard wrote in an essay called "The Transvaal":

We Englishmen came to this land . . . with "a high mission of truth and civilisation." . . . It is our mission to conquer and hold in subjection, not from thirst of conquest, but for the sake of law, justice, and order. (78–79, quoted in Katz 50)

And four years later, in *Cetywayo and His White Neighbours* (1881), he explained what the operation of law, justice, and order ought to do:

I cannot believe that the Almighty, who made both white and black, gave to one race the right or mission of exterminating or even of robbing or maltreating the other, and calling the process the advance of civilisation. It seems to me, that on only one condition, if at all, have we the right to take the black man's land; and that is, that we . . . allow no maltreatment of them . . . but on the contrary, do our best to elevate them, and wean them from savage customs. Otherwise the practice is surely indefensible. (270, quoted in Katz 146)

This is Haggard as public man, confident patronizing idealist, a founder-member, perhaps, of Conrad's International Society for the Suppression of

Savage Customs. In his fiction though, he has something else again to say. In two of his most celebrated early adventure novels, Haggard submits to some questioning the nature of civilization as it is experienced in colonial Africa. *King Solomon's Mines* (1885) is essentially an adventure story in which three white men search in pretty much unexplored territory for the lost brother of one of them—and for the fabled mines. They are accompanied on their search by a Zulu servant of great stature and dignity. After an eventful journey they stumble upon Kukuanaland, a hidden kingdom ruled by a tyrant (in "civilized" eyes at least) who is also a usurper. The Zulu servant turns out not to be Zulu but Kukuana, and the rightful heir to the throne. The white men help him regain his throne, but as they prepare to depart (having in the meantime discovered King Solomon's diamond mines), the new king says this to them:

> My people shall fight with the spear, and drink water, like their fore-fathers before them. I will have no praying-men to put fear of death into men's hearts, to stir them up against the king, and make a path for the white men who follow to run on. If a white man comes to my gates I will send him back; if a hundred come, I will push them back; if an army comes, I will make war on them with all my strength, and they shall not prevail against me. (306)

This desire to live in the traditional ways is, in the sense that we have been seeing the word used by Trollope and Haggard, savagery. But the effect of the passage is to introduce, if only for a fragment of the book, the modern relativist and comparative understanding of many different viable civilizations. The new king of Kukuanaland knows what the white man's civilization involves, and rejects it, and for a second or two within the narrative we are free to see his culture as different from, not inferior to, that of the whites.

A sequel to *King Solomon's Mines*, called *Allan Quatermain* (1887), ends with another statement of intent by the new (English) ruler of Zu-Vendis, another hidden African kingdom:

> I am convinced of the sacred duty that rests on me of preserving to this, on the whole, upright and generous-hearted people the blessings of comparative barbarism. . . . I cannot see that gunpowder, telegraphs, steam, daily newspapers, universal suffrage, &c, &c, have made mankind one whit the happier than they used to be, and I am certain that they have brought many evils in their train. I have no fancy for handing over this beautiful country to be torn and fought for by speculators, tourists, politicians and teachers, whose voice is as the voice of Babel . . . nor will I endow it with the greed, drunkenness, new diseases, gunpowder, and general demoral-

116

isation which chiefly mark the progress of civilisation amongst unsophisticated peoples. (276)

The notion of an Englishman as ruler of a hidden African nation is of a piece
with the paternalism of Haggard's public stance, and civilization is still
presented here as an absolute state. But at the same time "barbarism"
modulates into "unsophisticated peoples," and the catalog of the benefits
civilization will bring to the unsophisticated is yet more damning. Are these,
one might ask, Trollope's "good things"?[2]
    Allan Quatermain, the eponymous narrator of the story, considers this
question of civilization and savagery from a different viewpoint:

> It is a depressing conclusion, but in all essentials the savage and the child
> of civilisation are identical.
>   . . . [S]upposing for the sake of argument we divide ourselves into
> twenty parts, nineteen savage and one civilised, we must look to the
> nineteen savage portions of our nature if we would really understand
> ourselves, and not to the twentieth which, though so insignificant in
> reality, is spread all over the other nineteen, making them appear quite
> different from what they really are, as the blacking does a boot, or the
> veneer a table. (6)

This is the cynically reductive voice of the hunter who has observed over
many years interaction between black and white. The blacking simile has the
effect, whether consciously intended by Haggard or not, of forcing home for
the modern reader the conclusion that the only difference between black and
white is the color of their skin, a conclusion which is not quite Quatermain's. He implies throughout the passage that the black races are twenty-
twentieths savage; he does not allow even the momentary thought that they
might have their own surface of civilization. For him all humankind is
essentially savage, and under any kind of pressure the veneer of civilization
that automatically comes with white skin will be stripped under cover of a
puff of smoke, and white will behave precisely as black. Quatermain's
narrative proceeds to illustrate his conclusion. His group treks through
Kenya searching for Zu-Vendis and encounters the Masai. Unlike the
Zu-Vendi, the Masai are a real people with an ascertainable history, and the
record in the novel of the battles between colonists and Masai is one of
extreme violence and savagery on both sides (though the point of view
ensures that the white savagery is gloried in, and the black vilified). It is
characteristic of Haggard that as the whites examine the uncolonized land
they journey through, he should have Quatermain say (quite forgetting that
civilization is only skin-deep), "It is a glorious country, and only wants the

hand of civilized man to make it productive" (49), sounding the true note of the economic and human exploitation, echoing the voice of the conventional white settler, anxious to displace or exterminate the Masai and to plant his coffee bushes, that lies at the back of all that Trollope and Haggard wrote.

For Haggard, civilization remains a single recognizable social state, but he sees it from two separate angles. In his role as imperial apologist he considers civilization a moral instrument in the administration of Empire; but in his role as a novelist he suggests that in the colonial context civilization is a concept enabling the settler to justify doing as he will, and is primarily an instrument of corruption.

As the century ebbed, it became far more difficult to avoid relativity in the consideration of civilization, something made clear in two novels that first saw the light in 1899, Conrad's *Heart of Darkness* (1902) and Gissing's *The Crown of Life* (1899). They make a useful pairing because, though both are (among other things) interested in imperial undertakings in Africa and the function of civilization within them, Conrad concentrates on the periphery and Gissing on the metropole.

In order to test the new Imperialist ideals, in a novel written months before the outbreak of the Boer War, Gissing invents a character, Trafford Romaine, who is an amalgam of Rhodes, the visionary financial adventurer and prime minister of South Africa, and Milner, a chief administrative proponent of the new Imperialism, sent out as high commissioner to South Africa in 1897. We never hear Romaine speak, but one of the novel's central characters, Irene Derwent, is for a time deeply attracted to his ideas and discusses them widely. Gissing continuously undermines them, and as an example of his method, here is a snatch of dialogue between Irene and her aunts that refers to Romaine:

"He is immensely admired by some of our friends. . . . They compare him to the fighting heroes of our history."

"Indeed? . . . But the question is: Are those the qualities that we want nowadays. I admire Sir Walter Raleigh, but I should be sorry to see him, just as he was, playing an active part in our time."

"They say . . . that, but for such men, we may really become a mere nation of shopkeepers."

"Do they? But may we not fear that their ideal is simply a shopkeeper ready to shoot anyone who rivals him in trade? . . ."

"We are told . . . that England *must* expand."

"Probably. But the mere necessity of our case must not become our law. It won't do for a great people to say, 'Make room for us, and we promise to set you a fine example of civilisation; refuse to make room, and we'll blow your brains out!' One doubts the quality of the civilisation promised." (80)

As speech this sounds highly artificial, but it is effective rhetoric. The idealism that is still at this late stage inclined to identify civilization with the culture of late Victorian Britain is chipped away in this exchange to reveal the brutality at its base. Gissing also exposes the fraudulent use of civilization in the mouth of the economic exploiter:

> "I say that it's our duty to force our trade upon China. It's for China's good—can you deny that? A huge country packed with wretched barbarians! Our trade civilises them—can you deny it? It's our duty, as the leading Power of the world! Hundreds of millions of poor miserable barbarians. And . . . what else are the Russians, if you come to that? Can *they* civilise China? A filthy, ignorant nation, frozen into stupidity, and downtrodden by an Autocrat!" (254–55)

The ironies here for the modern reader are even more telling than for the late Victorian, and in the passage Gissing also embodies the arrogant assumption of racial superiority by the British, which lies behind their colonial enterprises, and which at its worst becomes pure Jingoism. When Irene Derwent is beginning to shape her views on Imperialism she asks a disciple of Romaine: "You take it for granted that our race is the finest fruit of civilisation?" (still assuming that civilization is sole and whole); she gets the remarkable reply, "Certainly. Don't you?"

Gissing identifies the newspapers as a primary source of this rabid racialism. The speaker here is becoming progressively more drunk:

> "I'm a journalist, Piers, and let me tell you that we English newspaper men have the destiny of the world in our hands. . . . We guard the national honour. Let any confounded foreigner insult England, and he has to reckon with *us*. A word from *us*, and it means war, Piers, glorious war, with triumphs for the race and for civilisation! England means civilisation; the other nations don't count. . . . You're not one of the muffs who want to keep England little and tame, are you? . . . I stand for England's honour, England's supremacy on sea and land. I st-and"—[and he falls over insensible]. (51–52)

"England means civilisation." Gissing, writing as it became clear that war was inevitable in South Africa, fears that the reverse may be becoming more accurate: that England means barbarism. Marlow, whose story forms the bulk of *Heart of Darkness*, does not, it seems, altogether agree.

119

Early on in his narrative Marlow echoes Haggard's suggestion about the rights and duties of colonists, though where Haggard had offered the imposition of "law, justice and order" as their justification, Marlow is characteristically vague about what redeems Imperialism:

The conquest of the earth, which mostly means the taking it away from those who have a different complexion or slightly flatter noses than ourselves, is not a pretty thing when you look into it too much. What redeems it is the idea only. An idea at the back of it; not a sentimental pretence but an idea; and an unselfish belief in the idea—something you can set up, and bow down before, and offer a sacrifice to. (50–51)

When Marlow sees on the wall of an office in the city of the whited sepulchre a map of the world "marked with all the colours of a rainbow," he comments that on it "there was a vast amount of red—good to see at any time, because one knows that some real work is done in there" (55).[3] Carlyle's reach down the nineteenth century was a long one, and from what follows in *Heart of Darkness* we can infer that in Marlow's justification for Britain's Empire there are mingled the ideas of work as moral activity and work as self-preservation, in a shape that was already familiar to metropolitan readers from Kipling's Indian stories. It is a question, however, much debated among critics as to whether there is identity or separation between Conrad and Marlow, and the question becomes acute if we consider this matter of civilization and savagery.

We might take as starting point Chinua Achebe's harsh but rational attack on the novel:

Africa as setting and backdrop which eliminates the African as human factor. Africa as a metaphysical battlefield devoid of all recognizable humanity, into which the wandering European enters at his peril. Can nobody see the preposterous and perverse arrogance in thus reducing Africa to the role of props for the break-up of one petty European mind? But that is not even the point. The real question is the dehumanization of Africa and Africans which this age-long attitude has fostered and continues to foster in the world.[4]

His argument is that all Africans are lumped together in the novel as undifferentiated savagery, and indeed since Conrad has chosen a European as first-person narrator, the criticism is in part inescapably true. But it might be argued that Conrad uses this perspective, as Haggard had, to say something about Western civilization.

When Marlow describes at the company station a file of Africans chained together, he says "They were called criminals, and the outraged law, like the bursting shells, had come to them, an insoluble mystery from the sea" (64). This is one of many sentences and paragraphs in the novel indicting the fatuousness (setting aside questions of morality) of imposing Western culture on an alien people; there is a quite savage sarcasm in "the outraged law." But what seems even more significant is the word "insoluble." Achebe points out that Conrad gives to Marlow's account of the wilderness and its people a string of such adjectives as "inscrutable," "incomprehensible," "unspeakable," "inexpressible," "even plain *mysterious*," and then suggests, with his own sarcasm, "When a writer while pretending to record scenes, incidents and their impact is in reality engaged in inducing hypnotic stupor in his readers through a bombardment of emotive words and other forms of trickery much more has to be at stake than stylistic felicity" (253).

But a hypnotic quality is in part just what Conrad is consciously after, in that it accurately reflects Marlow's sense of the repetitive and nonreferential quality of the jungle passing the ship; and further, the variation in terms of incomprehension represents precisely that bewilderment one has when placed in an environment which seems to bear no relationship to one's previous experience. In that environment (mostly physical here, though partly social and possibly psychological) one has no points of reference and is thrown back upon oneself, upon whatever there is to find in the self. Conrad thus, and Marlow perhaps, suggests that the Africans' response to the "outraged law" found operating at the "civilized" world of the seat of government and the main company station is an act of incomprehension parallel to Marlow's towards the "wilderness."

There is a paragraph on pages 96–98 of *Heart of Darkness* in which Marlow describes "the savage who was fireman" on the ship, and says that in engaging in such an activity he is "as edifying as . . . a dog in a parody of breeches and a feather hat walking on his hind legs," and that he "ought to have been clapping his hands and stamping his feet on the bank." In this simile Marlow clearly reduces the man to an animal, dehumanizes him. Further on Marlow says that this "savage" was "a thrall to strange witchcraft," the witchcraft of the ship's boiler, whose principles of operation he did not understand, but whose system of cause and effect he likened, in Marlow's conception of him, to an evil spirit inside the boiler which would blow them all up if it wasn't properly cared for. Marlow notes that the man has "an impromptu charm, made of rags, tied to his arm," presumably to help ward off hostile action by the evil spirit.

Some lines before this analysis of the fireman, Marlow discusses his failure to interpret the actions of the Africans passed on the bank during the voyage, and he notes the remote and disturbing sense of kinship he felt with those rituals, so different from the rituals of his own experience. He is sure that they were the rituals of "the night of the first ages," that change in man is a linear ascent from darkness into light, and he says that the mind of man contains "all the past as well as all the future." Motivating those songs and dances and gestures he recognizes there may be many things: "Joy, fear, sorrow, devotion, valour, rage—who can tell?—but truth—truth stripped of its cloak of time." And Marlow goes on to suggest that the only way a Western-civilized person can accept, assimilate the truth, and survive, or can reject it once it has been recognized, is to meet it "with his own true stuff—with his own inborn strength. Principles won't do. Acquisitions, clothes, pretty rags—rags that would fly off at the first good shake. No; you want a deliberate belief. An appeal to me in this fiendish row—is there? Very well; I hear; I admit, but I have a voice, too, and for good or evil mine is the speech that cannot be silenced."

Did Marlow consciously link his metaphorical representation of the moral, religious, ethical principles of Western civilization as "pretty rags" with the rag charm that the fireman wore on his arm? I think there can be no doubt that Conrad did. And the connection between advanced Western civilization and the night of the first ages, running counter to Marlow's assertion of unsilenceability, is enhanced a few lines further on when Marlow points out to one of his audience on the *Nellie* that he didn't join in with the Africans' dance because he "had to mess about with white-lead and strips of woollen blanket helping to put bandages on those leaky steam-pipes." This decoration with rags of the instrument of "strange witchcraft," the rickety vehicle of notation of difference between Marlow and savage, in terms of their relative civilized or uncivilized comprehension of the machine, serves imaginatively to lessen the gap between the two still further; which rags are more effective, Marlow's or the unnamed savage's? Any civilized person would naturally say Marlow's, until he or she considered the whole context of the paragraph, when some might be forced to say that neither is more effective than the other.

A further conclusion from these examples and from many other details of Marlow's early observations, is that Western civilization is a corruption, not an enlightenment. Marlow himself strives to assert his personal superiority to the unvisited Africans through the centuries of moral, cultural, and technological development; but he recognizes how few among the whites have learned anything from these centuries. The Eldorado Exploring Expedition is an example of this, and Marlow makes it clear that he understands what is at issue: "To tear treasure out of the bowels of the land was their desire, with no more moral purpose to the back of it than there is in burglars opening a safe" (87, though if one is to go by the Gould Concession in Conrad's *Nostromo* [1904], even when torn out with a moral purpose there is no guarantee of treasure's moral effect). Marlow judges them yet more harshly: "In a few days the Eldorado Expedition went into the patient wilderness, that closed upon it as the sea closes over a diver. Long afterwards the news came that all the donkeys were dead. I know nothing as to the fate of the less valuable animals. They, no doubt, like the rest of us, found what they deserved" (92).

When Marlow walks to the Central Station, he notes that the "population had cleared out a long time ago. Well, if a lot of mysterious niggers armed with all kinds of fearful weapons suddenly took to travelling on the road between Deal and Gravesend, catching the yokels right and left to carry heavy loads for them, I fancy every farm and cottage thereabouts would be get empty very soon" (70). As far as terminology is concerned, there is not a lot to choose between "niggers" and "yokels," though the former has become more offensive to more people since 1900; and the tendency of this hypothetical role reversal is to reduce both black and white to a common minimal level of humanity. Yokel, however, excludes Marlow's immediate audience (and their peers who would make up the majority of the readership of the novel)—the anonymous director of companies, lawyer, accountant, first narrator, representative figures of the Imperial power-structure. However, Marlow has a word for them, too. He tells them how the "overwhelming realities of this strange world of plants, and water, and silence . . . looked at you with a vengeful aspect. . . . I felt often its mysterious stillness watching me at my monkey tricks, just as it watches you fellows performing on your respective tight-ropes for—what is it? half-a-crown a tumble—" (93–94), and we are back with the fireman as a dog dressed up. All men, black and white, in the novel are reduced to the lowest common denominator. One of Marlow's audience interrupts him at this point with "Try to be civil, Marlow," and his choice of the word "civil" is directly appropriate. It is the power of the accretions of civilization that Marlow has effectively removed, just as Quatermain proposes in *Allan Quatermain*. It is arguable that when Kurtz wrote on his report the famous scrawled words "Exterminate all the brutes" he intended them to be applied to those who had sent him out to the Congo. Rather than a momentary revulsion from the culture he found himself inexorably slipping into, it is a judgment on those sentimental products of Western civilization who are so arrogantly confident of their superiority.

For Trollope, twenty years earlier and of a different generation, the concern was with civilizing the natives through work; for Marlow the more urgent concern has become the preservation of the self, irrespective of civilization. In part this reflects the growing fascination of late Victorians with the mechanisms of individual psychology, but it also in part embodies the growing sense that civilization was not after all a monolithic and enduring growth of Graeco-Roman origin that was reaching its climax in nineteenth-century Britain, but was beginning to appear multiple, fluctuating, precarious, and relative.

NOTES

Works are quoted from the following editions: Anthony Trollope, *South Africa*, 2 vols. (London: 1878); Rider Haggard, *The Days of My Life*, 2 vols. (London: 1926); *King Solomon's Mines* (Oxford: 1989); *Allan Quatermain* (London: 1887); George Gissing, *The Crown of Life* (London: 1899); Joseph Conrad, *Heart of Darkness*, Uniform Edition (London: 1923); Wendy R. Katz, *Rider Haggard and the Fiction of Empire* (Cambridge, Eng.: 1987).

1. Boswell said in 1772 that Johnson would not admit *civilization* into the fourth edition of his dictionary, though Boswell preferred it to *civility* as an opposite of *barbarity* (*OED*). This essay is the first published part of a long-term investigation into the history of the word.
2. It is instructive to compare these views of one of the imperialist Haggard's characters with those expressed in a book about Morocco by the arch anti-imperialist R.B. Cunningham Grahame:

   Guns, gin, powder, and shoddy cloths, dishonest dealing only too frequently, and flimsy manufactures which displace the fabrics woven by the women, new wants, new ways, and discontent with what they know, and no attempt to teach a proper comprehension of what they introduce; these are the blessings Europeans take to Eastern lands. (*Mogreb-el-Acksa*, 1898, quoted in *R B Cunningham Grahame*, by Cedric Watts [Boston: Twayne, 1983], 64)
3. On political maps, until quite recently, the British Empire was colored red or pink.
4. "An Image of Africa: Racism in Conrad's *Heart of Darkness*"; amended version in *Heart of Darkness*, ed. Robert Kimbrough (Norton Critical Edition, 3d edition, New York: 1988), 257.

KIPLING IN INDIA

Zohreh Sullivan, University of Illinois

Rudyard Kipling's writings about India are inscribed by the strains of late-nineteenth-century Empire as it reacted with increasing authoritarianism to post-Mutiny fears of the loss of British Raj. Kipling's hedged and tightly framed narratives, however, are also formal negotiations with his internalized, contradictory voices that both desired and feared India. In their ambivalence toward their embedded tales, these narratives suggest not only cracks in the ideological scaffolding that sustained the dreams of Empire, but also Kipling's more personal division between the contrary desires to explore and yet to contain the fearful and seductive stories of Indian life. As the voice of authority who provided the English with codes for surviving India, Kipling's work in the 1880s and 1890s provided alternative fictions of Empire that demythologized while it venerated the work of the English in other lands. His verse, journalism, and fiction generally form part of a larger colonial discourse—a historically produced language that however mediated, displaced, and symbolized, reproduces the varied voices of its uneasy and half-denied ideology, questions official structures, and raises the possibility of repressed and alternative re-readings of official moral and social stratification.

History, as Hannah Arendt has reminded us, needs myths and legends in order to contain that which otherwise could not bear examination. Historians, popular writers, and statesmen—those who create and formulate the stuff of legend, ideology, and mythology—are also those who shape group identity and shore up the idealizing fantasies of a community, thereby arming the culture against political reality.[1] Kipling provided his readers with legends and myths that reinforced British fears of the educated Indian, their adulation of the Pathan, and their faith in English schoolboy codes of manliness, sportsmanship, and honor ("The Head of the District" [in *Life's Handicap*, 1891], "His Chance in Life" [in *Plain Tales from the Hills*, 1888], "On the City Wall" [in *In Black and White*, 1890]). But Kipling also deflated such legends ("To Be Filed for Reference" [in *Plain Tales*], "The Phantom 'Rickshaw" [in *The Phantom 'Rickshaw*, 1890], "The Dream of Duncan

Parrenness" [in *Life's Handicap*]). Because of his understanding of the fragility of social contracts, of group psychology, of codes of conduct, and of breaking-points in the individual psyche, Noel Annan sees Kipling's place in the history of English ideas as analogous to that of Durkheim or Weber or Pareto—the European sociologists who revolutionized the modern study of society.

Kipling's representation of India, significant in the construction of a mythology of Imperialism, reflected both the real and the imaginary relationship between the British and their Indian subjects. What the community at home and in India chose to see in their enterprise was not what Kipling saw: they chose to see history and the work of Imperialism in the relatively totalized and glorified terms that Thomas Macaulay and Conrad's first narrator in *Heart of Darkness* (1902) saw it—jewels flashing in the night of time. But Kipling, like Durkheim and other sociologists, implied a disturbing analogy between private and public breakdown and cracks in the system. Robert Buchanan's famous attack on Kipling, "The Voice of the Hooligan," was characteristic of the complaint that instead of representing "the true spirit of our civilization" Kipling represents "the vulgarity, the brutality, the savagery" of "what the mob is thinking."[2] Buchanan wanted to see the Empire idealized, and in spite of Kipling's reputation as bard of Empire, his diverse and multi-voiced discourse, alternately ironic, factual, lyrical, and loving, resisted the single-voiced or monologic discourse of legislators and Imperial statesmen.

Yet at least one of the many "languages" and voices of Kipling served to preserve and petrify the "legend of Empire" as the eternally responsible parent who could be relied on to provide permanent support for the rest of the world. Although Kipling established what Arendt calls the "foundation legend" of Empire, the anxieties of his discourse suggest his awareness of the contradictions between the Imperial values celebrated in his tales for children and the political reality constructed by such men as Cecil Rhodes or Lord Curzon.[3] The construction of an idealized paternal imperialism that would protect childlike Indians may have been in part a compensation for the absence of protectors when Rudyard the child most needed them, for the violence of his own childhood deprivation of "reliable" parents, for his subjection to six years of abuse in what he called "The House of Desolation" and for the necessary repression of his "underground self"—the Indian child who was his Mowgli-self. The contradictions in Kipling's art, therefore, reflect those at the heart of the Imperial enterprise *and* in a particular family that provided the ideology responsible for producing the literature of fear and desire—fear of the "dark places of the earth" combined with the desire to master and control them; the need to master combined with the desire to rebel against authority.

126

Kipling's fiction undercuts the dominant ideology by using what Bakhtin calls a dialogic discourse—a method of representation that acknowledges its construction by history, politics, and culture and therefore realizes its diversity through conflicting and contradictory languages, styles, and characters. A man who praised Allah for giving him "two sides to my head," Kipling at his most lyrical (*Kim*, 1901) celebrates diversity by using a discourse apparently marked by unified, singleminded stances toward Empire that is then undermined by other decentering and "centrifugal" forces. Kim's numinous celebration of his journey on the multicolored, musical, and jewelled grand trunk road, "broad, smiling river of life," for instance, is made possible by his chosen, temporary identity as Indian and beloved *chela* to his Lama; but that position is later reversed by his confirmed identity as an Englishman whose "fettered soul" will see only a "great, grey, formless India." These contradictory images of India are repeated in a series of other historically inscribed contradictions, chief among which are Kim's desire to be loved by India as "little friend of all the World" and to be its master-sahib-imperialist.

Even a seemingly simple allegory of Empire like "Naboth" (in *Life's Handicap*) demonstrates the instability of apparent structures in Kipling's world. The story is about the gradual appropriation of an Englishman's garden (in India) by a poor street vendor toward whom he had made a thoughtlessly charitable gesture. The story, a nightmare of the lone English narrator victimized by ungrateful, insolent natives (a version of "The Strange Ride of Morrowbie Jukes" [in *The Phantom 'Rickshaw*]), ends with the following lines:

Naboth is gone now, and his hut is ploughed into its native mud with sweetmeats instead of salt for a sign that the place is accursed. I have built a summer-house to overlook the end of the garden, and it is as a fort on my frontier whence I guard my Empire.

I know exactly how Ahab felt. He has been shamefully misrepresented in the Scriptures. (*Life's Handicap* 289)

It is this final unsettling allusion to Ahab that is so characteristic of Kipling's mode, of the contradiction between the logic of his narrative sequence and its "secret" (to use Kermode's term), which the narrative has been at pains to repress. The story in Kings 21 tells of the annoyance of Ahab, king of Samaria, at discovering that Naboth the Jezreelite had a vineyard in Jezreel beside Ahab's palace. When Naboth refuses to accede to the king's demand to sell his land, Ahab sulks until his wife Jezebel has Naboth stoned to death. Ahab, the biblical colonialist, then takes possession of his coveted vineyard. What the biblical allusion does is to destabilize the monologic Imperialist

127

narrative voice and to raise questions repressed by the apparent allegory. Whereas Kipling's narrator dispels the reader's pity for the Indian and instead perceives the Englishman as victim to native ingratitude, the biblical story reverses the hierarchy, compelling the reader to question her early response to the problems of ownership, territoriality, and colonialism, to question the narrator's self-awareness, and to realize that the last sentence sets in motion an alternative story.

As product of a longing for "mine own people," Kipling's fiction swerves between the desire for a union with the Indian and the imperatives of the real and historic separation between the rulers and the ruled. The longed-for union is split between desire and dread: the idealized bond is realized in the native family structures that nurture Kim and Mowgli, and in Kipling's acknowledgment through the cryptic defense of subtitle (to *Life's Handicap*) and proverb that the native population encountered on the road to Delhi are all "my brothers" and "Mine Own People." But the dark underside to this idealized union is the terrifying bond seen in images of engulfment: opium dens ("The Gate of a Hundred Sorrows" [in *Plain Tales*]); brothels ("The City of Dreadful Night" [in *Life's Handicap*]); nightmare cities of the dead ("The Strange Ride of Morrowbie Jukes"); and, by extension, miscegenation ("To Be Filed for Reference," "Beyond the Pale" [in *Plain Tales*], "Without Benefit of Clergy" [in *Life's Handicap*]).

That conflict between desire and dread reveals itself in Kipling's tropology, his use of metaphor and metonymy. The metaphoric mode in Kipling blurs boundary distinctions by balancing forward movements toward discovery with backward movements controlled by fears and dreams that threaten disintegration. It risks the assertion of similarity between the self and Other, between the narrative voice and the blind face that cries and cannot wipe its eyes ("La Nuit Blanche"), and between the Englishman whose insomnia drives "the night into my head" and the Indians in the native cities of "Dreadful Night." The metonymic and ironic tropes in Kipling reinforce distance, contiguity, and denial of connection with the Other.[4] The all-knowing Kipling narrator of the early frame-tales uses irony as a pose, a defense against the terror of breakdown into an India of the mind. Kipling's Indian tales are pervaded by this dialogic battle between voices and tropes that threaten to undermine one another—the adventurer's desire for action, for example, countered by an opposite movement toward an ordered stasis. In *Kim*, for instance, the boy's apparent pursuit of adventure and freedom from familial and national obligations, signified by his initial choice of Lama as father and by his journey on the Grand Trunk Road into ever-wondrous images of India, is a forward movement that is contradicted by what we soon perceive to be the trajectory of his life—the rediscovery of emblems of Empire, of the Red Bull on a Green Field, of lost fathers, and of an inexplicably happy collection of parental figures whom he reterritorializes at play in the fields of a fantasy India.

These contradictory movements in Kipling's fiction match the conflicted attitudes of the colonizer toward the Other and toward competing ideas of Imperialism that preceded and followed the Mutiny. The vision of the united procession on the Grand Trunk Road in *Kim* is a decidedly pre–1857 construct, a throwback to a time before mutual distrust and suspicion led the British to adopt increasingly harsh authoritarian stances toward natives and to a time before the British felt justified in reneging on the 1858 proclamation promising that entry-level jobs in the government would be based on merit (Kiernan says that "statues of Queen Victoria were to multiply faster than jobs for Indians in the higher civil and military grades.")[5] Quite contrary to Kipling's Grand Trunk Road vision of a multicolored Empire marching toward the ultimate family reunion, the British continued to rely on an India divided against itself as the structure necessary to sustain their presence.

Although Kipling's own return to India in 1882 coincided with the growing agitation in Bengal that led in 1885 to the National Congress movement, most of his fiction appears to ignore native desires for independence. The anxieties in Kipling's earlier short stories, however, are provoked by such desires and the subsequent fear of loss. As the vernacular press in India was growing increasingly vocal in its opposition to British rule, as the voices of jingoism grew more intense in England, and as educated Indians began agitating for more power in government, the seventeen-year-old Kipling was beginning his seven hard years (1882–89) in India as a cub reporter for the *Civil and Military Gazette* in Lahore and for the *Pioneer* in Allahabad, trying to shore gravel against the tides of political change— against the Ilbert Bill and Gladstone—by writing essays demonstrating the inability of Indians to govern themselves. The country was for him too large, too diverse to be governed by anything other than a united Empire. Although his articles were fearful of inefficiency and destruction if Indians governed themselves, the bulk of the fiction he wrote during these years was about the inefficiency and self-destructiveness of the English. Madness, alcoholism, self-doubt, and suicide haunt the characters who live in the plains and hills of his early volumes of stories.

Kipling's greatest defense against the horror of self-loss and disintegration lay in his discovery of a narrative stance at once distant, ironic, hedged, and monologic—a voice that reminds us of his characteristic fictional narrator's voice so persistently contradicted by the polyphonic voices of the embedded tale and by the diverse voices of native India. That single-minded and easily caricatured voice belongs to the daytime Kipling, who staunchly defends Imperial values and institutions; but what distinguishes Kipling's work is the presence of other voices, dissonant and self-contradictory, that war with the monologic voice and signify Kipling's ambivalences toward the dominant preoccupation of his early and middle fiction—how to survive and parent India. Kipling's fiction of the 1880s and 1890s moves forward toward resolving the problem of how to exist in the face of the potential political and personal loss of India ("The Bridge Builders" [in *The Day's Work*, 1898]), and backward to a primal loss ("Baa, Baa, Black Sheep" [in *Wee Willie Winkie*, 1890], "The Strange Ride of Morrowbie Jukes") that reacts to such a possibility with hysterical defensiveness and denial.

The paradigmatic figure of these contradictory movements in Kipling's fiction is sight and sightedness. The stories are full of images of perception—ghostly, half-eaten, mutilated faces ("In the Same Boat" [in *A Diversity of Creatures*, 1917]), blind men (*The Light that Failed* [1890]), blind women ("They" [1904], "The Wish House" [in *Debits and Credits*, 1926]), nonhuman creatures whose ghastly eyes are "sightless—white, in sockets as white as scraped bone, and blind" ("A Matter of Fact" [in *Many Inventions*, 1893]), and the twin figures of Kim, who *sees* in brilliant colors a painterly India, and his Lama, who refuses to see and who denies the significance of sight. But sight is as terrifying as blindness, and variations on the problems of too much perception resonate through his poetry and fiction: the fear of vision and revelation that results in the destruction of the first microscope ("The Eye of Allah" [in *Debits and Credits*]) in a thirteenth-century monastery, the literalizing of the metaphoric fear that "truth blinds the perceiver" ("A Matter of Fact"), and the suggestion that those granted special sight are also damned to madness and death ("The Phantom 'Rickshaw" and others). The fear of seeing the self as other occurs in stories of doubles ("The Dream of Duncan Parrenness") and most starkly in his poem "The Mother's Son" (1932):

> I have a dream—a dreadful dream—
> A dream that is never done,
> I watch a man go out of his mind,
> And he is My Mother's Son.

But Kipling's project in life was to overcome such destruction. The absence and lack suggested by blindness he therefore meets by the opposing urge to see all—even the forbidden city streets of nighttime Lahore—and to confront the fear of immobility by obsessive traveling in both fiction and life. In his fiction, Kipling became the supreme framer of tales in which a survivor tells of the dismemberment, madness, or death of another who ventured where frame-narrators fear to tread. Kipling's ironic and metonymic layering of narrators and stories refracts the connections among himself, his narrator, and the embedded tale, thereby repressing the metaphoric figures associated with loss of self by displacing them into a contained and embedded tale. His stories, in other words, commit acts of official transgression by centralizing what Imperial society had marginalized—the failure, the suicide, the syphilitic soldier, the rebel, the Indian child, the Indian woman, the dying, and the dead.

Elaborate tropes, the coyness of public school "in" languages, and the armored style of military language conceal internalized disorders that seem inexplicable. Mysterious suicides are explained only by "had a touch of the sun, I fancy" ("At the End of the Passage"), or syphilitic fits interpreted by doctors as "Locomotus attacks us" ("Love-o' Women" [in *Many Inventions*]), or madness diagnosed as "hydrophobia" ("The Mark of the Beast" [in *Life's Handicap*]). The fear in most of these stories is of internal rather than external "powers of darkness." The epigraph to "The Phantom 'Rickshaw" returns as a hymn sung by one of the men in "At the End of the Passage":

> If in the night I sleepless lie,
> My soul with sacred thoughts supply;
> May no ill dreams disturb my rest,
>
> . . . .
> Or powers of darkness me molest! (*Life's Handicap* 146)

Both ill dreams and the powers of darkness drive the English into early graves. But the men in "At the End of the Passage" envy the suicide and prefer his death to their endless assault by isolation, heat, sleeplessness, stress, overwork, and monotony. Hummil, the man who has reached the end of his passage, dies of fear, haunted and hunted in his sleep by "a blind face that cries and can't wipe its eyes, a blind face that chases him down corridors." Kipling clearly understands his disorder as fear of regression: "He had slept back into terrified childhood . . . he was flung from the fear of a strong man to the fright of a child as his nerves gathered sense or were dulled" (*Life's Handicap* 151). Hummil's servant believes that his master has descended into "the Dark Places, and there has been caught," which is another way of describing a psychotic breakdown, or regression into the terrors of an infantile past and into "the echoing desolation" of India as a projection of the anxieties of that past. This is a paradigmatic story of colonial fear of boundary loss in India, the fear of loss of control, of "slipping backwards" into childhood, darkness, alcoholism, passion, quick-sand, or an abyss. Whereas most of the stories of the 1880s are controlled by a frame-narrator who carries us down only to return us to daytime safety, this story has no mediating frame to separate the audience from the inevitability of dissolution that seems to await every man who is not a frame-narrator in India.

The problem of survival in colonial India, therefore, realizes itself in psychic and relatively ahistorical chains that cross specifically historical chains of meaning. Kipling maps his human failures—the blind and the mutilated—by certain recurring tropes charged with the fear of slipping backwards into a past that might once again be lost. Chief among these are nightmare images of wounds or scars in the geographic landscape that are metonymically related to corresponding bodily mutilations. The gaping hole of the village of the dead who did not die, for instance, traps the unwary Morrowbie Jukes, who finds himself at the mercy of a native with a scar for which he has, sometime in the past, been responsible. The one-eyed priest in "Bubbling Well Road," who carries "burnt between his brows" the scar of the torturer, guards an eerie, haunted well that is "a black gap in the ground" (*Life's Handicap* 266). And the well into which Little Tobrah pushes his starving sister is one of the silent spaces in India that speaks not only of its own loss, but also of the endlessly postponed and displaced negation that signifies Kipling's search for the forever evasive and absent "cause" that is forever a lack. The gaps in the ground and in the priest's face are metonymies that, by definition, are not only endless, but also "constantly confused, traversed by other chains . . . a crossroads, a vertigo,"[6] that serve also as historic metonyms of the black hole where the English in India (particularly after 1857) might fear to meet their living death. These powerful early stories, marked by colonial and racial fears, undercut the illusions of common inheritance and "grandeur of race" (Charles Wentworth Dilke's phrase) that served Imperial power, not as historical facts, but as "a much-needed guide . . . the only reliable link in a boundless space."[7]

The fear of loss, irrational yet repeated and persistent, appears already in Kipling's early journalism, where he describes journeys into grotesque geographic spaces that prefigure the later Conradian heart of darkness. Most striking for their overdetermination are his infernal descent into "The City of Dreadful Night," and his visit to the Gau-Mukh ("cow's mouth") in Chitor (*Letters of Marque* [1891], 11.88–99), which he recognizes and names as "uncanny" and which then reappears as the center of darkness in *The Naulahka* (1892). In the *Letters of Marque*, Kipling feels he has "done a great wrong in trespassing into the very heart and soul of all Chitor" (95). He has literally slipped into a sexualized landscape of absence split between a tower and a gaping watery hole that contains "the loathsome Emblem of Creation" (94), with flowers and rice around it. The terror he feels at seeing the phallus (the lingam) underwater enclosed by incense, vegetation, and "oozing," "trickling," and "slimy walls" makes Kipling feel he has been led "two thousand years away from his own century . . . into a trap" and into the fear that he "would fall off the polished stones into the stinking tank, or that the Gau-Mukh would continue to pour water until the tank rose up and swamped him, or that some of the stone slabs would fall forward and crush him flat" (95). Sweating with what he terms "childish fear," he decides "it was absurd. . . . Yet there was something uncanny about it all" (96). Although this piece was written about twenty years before Freud's essay on "The Uncanny" (1919), it is remarkable for its precise anticipation of the term: what the Englishman sees in the phallus dismembered and enclosed by water is the return of the repressed fear of absence, castration, and self-loss. That fear is replayed in Kipling's dream story "The Brushwood Boy" (in *The Day's Work*), where the boy dreams of a house containing a fearful Sick Thing, an It, whose head threatens to fall off, and in "The Man Who Would Be King," where Dravot's decapitated head and Carnehan's crucified body and stigmata are signs not only of their journey into a forbidden Indian darkness, but also of the penalty for crossing a sexual boundary.

Kipling wrote in a world charged with the desire for possession and ownership: "Expansion is everything," said his friend Cecil Rhodes, "these stars . . . these vast worlds . . . I would annex the planets if I could."[8] The central event of Kipling's Indian stories is the act of transgression of an inner circle, a family, or a world that is contained, questioned, and sometimes repressed. The violation of territorial, sexual, and moral boundaries consistently recurs in stories where the act is sanctified by religiosity ("The Man Who Would Be King") or sentimentalized by making the transgressor a child (*The Jungle Books* [1894, 1895] and *Kim*), or distanced into dream ("The Brushwood Boy") and ironized by a frame-narrator whose language and stance attempt to neutralize its impact ("Beyond the Pale," "Naboth," "The Phantom 'Rickshaw").

Kipling's most powerful early Indian stories, versions of his longed-for imaginary union, tell of Englishmen who attempt to violate divisions between the colonizer and colonized by learning too much about nighttime India, by consorting with native women and thereby succumbing to the seduction of drugs, alcohol, and the Other. Such surrender might lead to madness and death—or, to writing the ultimate book about India. "To Be Filed for Reference" is the last story in *Plain Tales from the Hills*, the last word (at least in 1888) on the colonizer's transgression into native India. Here the Oxford scholar McIntosh Jellaludin is evidence of the cost of transgression—now perpetually drunk and "past redemption," cared for by his native "wife," he calls the narrator to him as he dies. The book that is his life's work, "my only baby" (241), which he wishes to bequeath to the narrator, is called "the Book of Mother Maturin"; and while the narrator begs to be dissociated from its authorship ("remember . . . that McIntosh Jellaludin and not I myself wrote the book of Mother Maturin" [241]) we know that Kipling's most ambitious project in India was to be an epic novel called *Mother Maturin* about a Eurasian woman "not one bit nice or proper" that dealt with "the unutterable horrors of lower class Eurasian and native life as they exist outside reports."[9] It is this slippage in Kipling's Indian fiction between himself and the Other, himself in disguise as the Other, himself in love with the Other, yet not the Other but instead a cold, armored, and distant narrator that makes his fiction so characteristic of his metonymic mode of narrative—one that can never quite locate its center, or name its unnamable desire.

That which is named as fearsome, undesirable, or objectionable in one part of his work, for instance, often returns (and this is a central characteristic of the metonymic mode) to haunt another part of his fiction as ambivalent object of desire. The series of articles, for instance, that Kipling wrote about Calcutta ("The City of Dreadful Night," 1888) charts an infernal spiral descent from the surface "whited sepulchre" (*From Sea to Sea* 196) to the Calcutta Stink ("the essence of long-neglected abominations" [224]), down through labyrinthine layers of brothels, "from love by natural sequence to death" (243). The infernal descent reserves the worst layer for the last—the sight of European women, once wives of British soldiers, now working as prostitutes. The section ends with the reporter's urge to climb the spire of a church and shout: "O true believers! Decency is a fraud and a sham. There is nothing clean or pure or wholesome under the stars, and we are all going to perdition together. Amen!" (328).

Here and in the story of the same name, Kipling expresses the wish to climb a spire, to ascend to a position from which ("like Zola") to frame a view. In the earlier "City of Dreadful Night" (1885) story about a sleepless night in Lahore, he does what he wishes to do in the later piece—climb the top of a minaret and then look down on the city: the characteristic stance of the narrator who finds ways to distance himself from immersion in the destructive element. In each version of the tale, the heart of darkness is death—the Eurasian prostitute is the death of illusion, and the all-night walk through Lahore begins and ends in death. The opening paragraph of the story is a familiar vision of India—"a disused Mahomedan burial-ground, where the jawless skulls and rough-butted shank-bones, heartlessly exposed by the July rains, glimmered like mother o'pearl on the rain-channelled soil." The narrator, who has begun by complaining that the dense, wet heat "prevented all hope of sleep," ends the story and the journey through yet another night of insomnia by seeing a double, a nightmare image of yet another victim of that heat—the corpse of a woman of whom someone says: "She died at midnight from the heat."

The native woman in Kipling's tales, though not always silent or dead, is a presence whose evocation of desire and fear connects with the personal and historical tropes of family inscribed in the discourse of Imperialism. In Kipling's love stories of miscegenation, where the Englishman loves or lives with a native woman, the woman's very presence gives a voice to an otherwise marginalized gender and culture. But her existence is significantly and often ironically undercut and minimized by the narrator, by the imperial society that the voice of the narrator represents, and by history. In "Beyond the Pale," a story of an ill-fated affair between the Englishman Trejago and a beautiful fifteen-year-old Indian widow, both the narrative stance and the structure seem at pains to repress the unknowable problem at the heart of the labyrinth of dead-end streets—the problem of native life. The narrator has so mastered knowledge of Indian life that he can safely announce "much that is written about Oriental passion and impulsiveness is exaggerated and compiled at second-hand, but a little of it is true" (131). The story that ends with the gruesome vision of the girl's arms held out in the moonlight— "both hands had been cut off at the wrists, and the stumps were nearly healed" (131)—is framed by the dominant voice of the Imperialist who rewrites a story of native tragedy (the girl is all but buried alive and mutilated) into a Victorian "advice" story designed to help young Englishmen in the tropics: "This is the story of a man who wilfully stepped beyond the safe limits of decent everyday society, and paid for it heavily. He took too deep an interest in native life, but he will never do so again." This, the first line of the story, is nicely balanced against the last line of the story that chronicles the effects of his affair and near castration: "But Trejago pays his calls regularly, and is reckoned a very decent sort of man. There is nothing peculiar about him, except a slight stiffness, caused by a riding-strain, in the right leg" (132).

Although surface meaning in this story is produced by a series of implied opposites understood in terms of gender, culture, and geography (its setting is a "trap" in the "heart of the City"—a space that "ends in a dead-wall pierced by one grated window," the native girl is a "child" with "little feet, light as marigold flowers, that could lie in the palm of man's one hand"), the repressed meaning is produced by what is absent, the gap that gives the narrator a momentary glimpse into native life but that gets "walled up" before he can understand what he has seen (132). Kipling's romance reenacts the fantasy family affair between India and England and blames for the end of the affair the "unreasonable" jealousy of the Indian who refuses to share the devotion of the Englishman toward his club and culture. Though it uses the cliché of "falling in love" in order to reenforce a fantasy of equality and to efface the power relationship between the ruler and the ruled that dooms such a dalliance, Kipling's narrator plays with the inequality throughout: Bisesa, he says, was "an endless delight," as "ignorant as a bird" whose reactions to his daytime affairs "seemed quite unreasonably disturbed"; but it is she who determines the end of the affair, it is she who swears that no harm will come to him no matter what her fate.

The problem of fixing meaning in this story is created by confusions of gender, culture, knowledge, and geography. Finally, she—the child with roseleaf hands—is victimized not only by the affair, but by the uncle with whom she lives. At Trejago's final visit "someone in the room grunted like a wild beast, and something sharp—knife, sword, or spear—thrust at Trejago in his boorka." At the end, the slightly wounded Englishman rages and shouts "like a madman between those pitiless walls" (132); the native girl is walled forever into a borderline space between the human and nonhuman, life and death, culture and chaos—embodying the absence, irrationality, darkness, beastliness, the castration that the European fears. Her missing hands repeat the trope of other absences—the mysterious house that cannot be found, the missing front door, the missing effaced cause of the tragedy (serialized conquests of the girl by the heirarchic Indian tradition, by father, husband, uncle, and English lover, homologous to the serialized conquests of India by countless others). Trejago and the narrator repeat the trope through their lack of final knowledge of what the story means: he "does not know" what became of the girl, he "does not know" where the front of the house of his beloved is, and he cannot get "poor little Bisesa" back again. "He has lost her in the City where each man's house is as guarded and as unknowable as the grave; and the grating that opens into Amir Nath's Gully has been walled up."

Desire and fear in Kipling's split discourses about India yearn toward and reject the heart of their darkness that lies in the labyrinthine alleys of nighttime Lahore. Kipling's stories of India struggle with that nighttime desire for knowledge in competition with the opposing need for the surface stability of its daytime trope—the familial enclaves of the inner circle, the club, the home, and the Empire. His negotiation of those desires in his early fiction produces tensions between the privileged, dominant, and dispassionate position of the frame-narrator and the more inflected positions contained within the embedded and enigmatic tales. Those embedded tales often release voices that oppose the dominant voice of hegemony. It is Lispeth, the hill girl, that the reader supports, and not her English lover or the English missionaries who try to educate her into civilization. Sometimes that dominant voice is compromised by irony, sometimes by an elegiac compassion for crazy adventurers who would be kings and who go beyond the pale into forbidden India; but such men meet with mutilation, decapitation, or crucifixion. The survivors are those who are supported by fantasy families: Mowgli the "Frog," amphibious child of jungle and city, of humans and animals, can live in the Jungle and not be of the Jungle; Kim the orphan, loved by a host of multicolored parental surrogates, can be "little Friend to all the world" and master; and Adam, son of Policeman Strickland, can earn the "double wisdom" necessary to survive Imperial India with humanity and dignity. Other Englishmen are allowed to survive India because they are

"Lispeth," bas relief by Lockwood Kipling, from Rudyard Kipling's *Plain Tales from the Hills* (1900).

THE ENDS OF THE EARTH: 1876–1918

granted the privilege by ancient Hindu Gods who, in reversal of the colonial hierarchy, are represented as the real controllers of "the day's work" when they allow bridge-builders to realize the rewards of their labor.

But all these certainties are threatened by the persistence of Kipling's original trope of uncontrollable terror: in spite of the Englishman's hard work, Naboth will invade the garden when the Englishman least expects it; the Indian gods will cause floods to wreak havoc on the work of Empire; Morrowbie Jukes will lose his head, start shooting howling dogs at midnight, and fall into the city of the dead from which there is no escape; the ghost of Mrs Wessington will return to haunt Pansay to death; and the unnamed "Boy" will blow his brains out for no nameable reason. Kipling's writings about India, produced at a particular point in history, implicate the reader in some of the recurring controversies about the human and psychic cost of maintaining the Indian Empire. The issues his stories address are those raised by Disraeli in his famous 1872 speech in the Crystal Palace: "The issue is . . . whether you will be a great country,—an Imperial country—a country where your sons, when they rise, rise to paramount positions, and obtain not merely the esteem of their countrymen, but command the respect of the world." But unlike Disraeli's, Kipling's Imperial ideology is disrupted by doubts and anxieties, by an oppositional, slippery, and unstable stance on knowledge reflecting voices that, by interrogating the monologic utterances of Imperialism, direct their irony at man's all too human and suspect need to control and conquer nature and the wider world.

NOTES

Quotations from Kipling's works are from the following editions: *Plain Tales from the Hills*, ed. Andrew Rutherford (Oxford: 1987); *Life's Handicap*, ed. A.O.J. Cockshut (Oxford: 1987); *The Man Who Would Be King and Other Stories*, ed. Louis L. Cornell (Oxford: 1987); "The City of Dreadful Night" in *From Sea to Sea* (Garden City, NY: 1932); *From Sea to Sea: Letters of Marque* (Garden City, NY: 1932).

1. Lewis Wurgaft, *The Imperial Imagination* (Middletown: 1983), x–xii.
2. Elliot L. Gilbert, *Kipling and the Critics* (New York: 1965), 20–32.
3. Hannah Arendt, *The Origins of Totalitarianism* (New York: 1973), 209.
4. My terminology is indebted to, among others, Hayden White (*Tropics of Discourse*, 1978)—who sees the process of understanding itself as tropological, a process analogous to Piaget's study of cognition, a movement from the unfamiliar and "uncanny" (what Lacan would call the Imaginary), to the more familiar, ironic, and Real; he draws further analogies between Freud's four mechanisms of dreamwork and the four "master tropes" of figuration—metaphor, metonymy,

synechdoche, and irony. My ideas of the text and the subject are influenced primarily by Mikhail Bakhtin (*The Dialogic Imagination*, 1981) and Fredric Jameson (*The Political Unconscious*, 1981), who historicize literature and the self by recognizing their inscription by the many and often contradictory voices of culture and ideology.
5. V.G. Kiernan, *The Lords of Human Kind* (New York: 1986), 51.
6. Jean-François Lyotard, "The Tensor," *Oxford Literary Review* 7 (1985), 30.
7. Arendt, 182.
8. Ibid., 124.
9. Norman Page, *A Kipling Companion* (London: 1984), 158.

"THESE PROBLEMATIC SHORES": ROBERT LOUIS STEVENSON IN THE SOUTH SEAS

Barry Menikoff, University of Hawaii

British literature offers few pictures more exotic or romantic than that of Robert Louis Stevenson—brilliant and charismatic, yet condemned to self-exile and early death on a remote island in the Pacific Ocean. From these minimalist details, a series of portraits of the artist were constructed over the years that augmented and embellished both the exoticism and the romance. In a way, nothing that Stevenson wrote could compete with the stories that were written about him, fictions created not only by writers "on the trail" of Stevenson but by his own family and friends, all of whom had an interest in the novelist's reputation, not to mention a share in the literary estate. Without question, the adulation visited upon Stevenson during the last years of his life and immediately after his death was a factor in the reaction to his work that set in shortly after the First World War. Henry James, who knew Stevenson well, saw this clearly as early as 1901, as he wrote to Graham Balfour, Stevenson's cousin, upon reading the official biography: "You have made him—everything has made him—too personally celebrated for his literary legacy."[1] Added to the personal factor, librarians and educationists solidified the classification of *Treasure Island* (1883) and *Kidnapped* (1886) as children's literature, and academics ceased to read Stevenson altogether. In time, the scope and variety of his writing was forgotten, and what debate there was focused on whether Stevenson's important contribution was to the essay form or the novel. Edmund Gosse, for example, whose reputation as a man of letters benefited from his friendship with Stevenson, insisted upon the superiority of the essayist in his entry on Stevenson for the eleventh edition of the *Encyclopaedia Britannica* (1911).

This is not the place to review that false argument, except to say that it seriously discounted work that did not fall into either of those categories. In particular, it paid no critical attention to the travel writing that Stevenson produced over the course of his literary career. *An Inland Voyage* (1878) and *Travels with a Donkey* (1879), his earliest published books, were perennial favorites, the latter enjoying a devoted following among lovers of France and the region of the Cevennes. Later, when Stevenson crossed the Atlantic and traveled over the North American continent as an "amateur emigrant" he developed an entirely new group of admirers, those residents of the United States who enjoyed the sharp and even raw vignettes of life among the newer classes of Americans as seen by an observer who, unlike many of his earlier countrymen, was more sympathetic than antagonistic to his linguistic cousins. Finally, after Stevenson married an American woman in San Francisco and "squatted" in an abandoned mining camp in Silverado, the resultant volume, *The Silverado Squatters* (1883), achieved the status of a cult classic among Californians. In the interim, Stevenson had written essays on Monterey and San Francisco as well as extended reflections on life past and present in the Napa Valley.

The importance of this writing cannot be overestimated, if only because it serves to remind readers that Stevenson as a professional writer was less interested in genres and categories than he was in subject matter and material. Just as he theorized about the nature and art of fiction, he was equally reflective about technical and theoretical issues concerning the nature and practice of travel writing. Travel for Stevenson was a process of personal and cultural exploration, at times even a condition of experience. And it was always bound up with history. Thus in *Travels with a Donkey* Stevenson read widely in French sources on the wars in the Cevennes and incorporated that reading in his text. The process was continued in California, where he carried Bancroft's volumes (purchased in New York) for use as a fundamental reference.[2] Authorities were essential because his work depended upon historical information. Thus, Stevenson's eyewitness account of the ethnic character of modern San Francisco would be balanced or supported by a historical summary of the development of the city as the "new" Pacific capital and its displacement of the "old" capital at Monterey. The effect of Stevenson's method was to make him a "travel" writer whose work was always deeper, or at least less superficial, than readers were ever aware of. Stevenson invariably concealed his sources—at the most he would provide sly hints of their existence—and he developed from early on a method by which he could incorporate documentary information within a narrative that was both readable and personal, one that appeared to be nothing more than one traveler's observation and experience. The result has been the uniformly high status of *Travels with a Donkey*, *The Amateur Emigrant* (1895), and *The Silverado Squatters*.

But Stevenson's final volume of travel pieces faced a very different reception and history. He had matured as a writer, and his vision had broadened and deepened. Also, the material he was working with was inherently more interesting. Yet, paradoxically, it was these very factors that consigned his articles on the South Seas to virtual oblivion. They were a dismal failure at the time of their newspaper publication, and they fared no better in book form. There was a general consensus that *In the South Seas*, published posthumously in 1896, was a complete departure from Stevenson's habitual style and method, and it would have been better had he never written it. Of course there were dissenters, occasional readers shrewd enough to insist upon the authenticity and brilliance of Stevenson's discourse on various isolated Pacific archipelagoes. But these readers were usually people with firsthand experience of the islands, rather than literary critics or historians, and their judgment was perforce discounted. Thus one of his most original and imaginative experiments, his attempt to formally convey the complexity and diversity of peoples of color to an English and American audience, was destined to an ignominious end: it was modified, censored, and finally cancelled in its serial form. Stevenson never had the opportunity to reconstruct his definitive text on the cultures of the Pacific islands in the final quarter of the nineteenth century.

In 1887 Stevenson left England for the United States. He first spent a winter in a sanatarium at Saranac Lake, during the historic New York blizzard of 1888. Later, while visiting in Manasquan, New Jersey, he received a telegram from his wife that a yacht was available for hire from Dr. Samuel Merritt of Oakland. Stevenson telegraphed back to accept the offer, and he left immediately for California. The quickness with which he jumped at the telegram ought not to suggest impulsiveness. If the decision was abrupt it was not capricious. Stevenson's poor health, which had always been a dominating factor in his life, motivated him to make the trip. Turning his back on England, and in a sense on the United States, the South Seas offered a tantalizing alternative to perpetual respiratory and pulmonary sickness. And, in fact, Stevenson did feel better in the Pacific. But the islands turned out to be more than a simple restorative: they possessed an enormous intellectual and imaginative attraction for the Scots writer. Stevenson loved the sea, and he read widely in the explorers and hydrographers who had traveled and charted the Pacific Ocean. He was mentally prepared for his journey in ways that those who knew him were ignorant of, and that he himself was probably unaware of. The *Casco* gave him the opportunity to make good on a dream which may have been as inchoate as it was compelling—the dream of renewed life in a new world.

Stevenson's trip was made possible by his agreement with S.S. McClure to write a series of articles, or "letters," for newspaper syndication. The payment for these would underwrite the expenses of chartering the schooner. The project had a long incubation period.[3] On December 8, 1888, McClure wrote to Stevenson that he had contracted with the New York *Sun* to publish up to fifty letters at $200 apiece, a contract he considered "singularly fortunate" after he discovered it would have been almost "impossible" to secure widespread newspaper syndication—"editors in general regarding this subject as too remote."[4] Stevenson's reputation notwithstanding, even at this early stage of negotiation McClure was forced to acknowledge the problematic nature of the proposed material. Additionally, he had to contend with Stevenson's abhorrence of writing for deadlines. On January 10, 1889, McClure again wrote to Stevenson about the $10,000 agreement for the newspaper articles: "Of course you understand that the contracts are made on precisely the terms which you arranged. You are not bound to write one article nor any special time nor to finish them at any special time; you are absolutely free" (Beinecke). McClure worked out an agreement that appeared to give Stevenson the best of two worlds: he would receive substantial payment for the letters, and he would not be held to an impossible schedule. Needless to say, deadlines were still imposed, and Stevenson found himself turning out massive amounts of copy within relatively tight time constraints. And, for an added twist, the *Sun* proved to be no more sophisticated with respect to their appreciation for Stevenson's subject than McClure's less cosmopolitan newspapers.

McClure had Stevenson's first thirteen letters bound up in a small red book, with chapter headings for each letter and the title *The South Seas* printed on the cover.[5] The *Sun* was unhappy with this volume. As McClure wrote to Stevenson on January 31, 1891: [the *Sun*] "refused to regard these chapters as being your letters, contending that they were not furnished letters at all, but simply advance sheets of a book of travels, and refusing utterly to carry out the contract which was made with them on the ground that these chapters were not what they bought" (Beinecke). McClure argued that the book publication was a device to secure accurate proofs. The *Sun* was not persuaded. They "objected to the fact that the letters did not come as letters are supposed to come. They were not a correspondence from the South Seas, they were not dated and they contended that in no way did the matter in the first series fulfill the definition of the word 'letter,' as used in newspaper correspondence." McClure was finally successful in getting the editors to accept the material, with the proviso that the letters would be syndicated rather than remain the exclusive publication of the New York paper.

The newspaper editors were not alone in their concern over the articles. Sidney Colvin, Stevenson's close friend and de facto literary agent, was adamant in his opposition to the South Seas writing. Stevenson's letters to Colvin are repeated efforts to explain and justify his new aims and methods, and to get Colvin to soften if not suspend his criticism. But the *Sun*'s initial rejection of the slim red volume was all that Colvin needed to justify his attitude toward "the South Seas book." As he wrote to Charles Baxter, Stevenson's closest friend, on February 4, 1891, the "stuff is not what [they] asked for; that is to say, not letters of incident and experience, hot & hot from the scenes described, but only the advance sheets of a book, and rather a dull book at that." On April 11, after installments of the series appeared in *Black and White* in London, Colvin again wrote to Baxter:

This South Seas Islands book, as you will have realized if you have read the chapters in Black & White, is the very devil for dulness, confusion, and ineffective,—even incompetent workmanship: and all because he would insist in laying aside personality, incident, adventure, and humour . . . and stuffing it with half-digested information and bewildering allusions to things no reader either knows or wants to know about. (Beinecke)

But Colvin's ignorant, even philistine attitude was not Stevenson's. From the beginning the novelist exhibited his openness to an experience that was to prove profoundly revelatory.

> Yacht Casco
> Anaho Bay
> Nouka Hiva
> Marquesas Islands

My dear Colvin,

From this somewhat (ahem) out of the way place, I write to say how d'ye do. It is all a swindle: I chose these isles as having the most beastly population, and they are far better, and far more civilised than me. I know one old chief Koo-amua, a great cannibal in his day, who ate his enemies raw as he walked home from killing 'em, and he is a perfect gentleman and exceedingly amiable and simple minded: no fool, though.

The climate is delightful, and the harbour where we lie one of the loveliest spots imaginable. Yesterday evening we had near a score natives on board: lovely parties. We have a native God: very rare now; Very Rare and Equally Absurd to View. This sort of work is not favourable to correspondence. It takes me all the little strength I have to go about and see, and then come home and note, the strangenesses around us.

I shouldn't wonder if there came trouble here someday all the same: I could name a nation that is not beloved in certain islands—and it does not know it! Strange: like ourselves, perhaps, in India! Love to all and much to yourself R.L.S.[6]

Stevenson can barely contain his delight at this brave (or strange) new world, and it is ironic that within two years he finds himself chastising Colvin for the latter's "damnatory" comments about the *Letters* while they are running in the newspapers.

Stevenson repeatedly insisted to Colvin that his South Seas *Letters* were to be viewed as nothing more that work-in-progress. His main objective was the construction of a book: "Can I find no form of words which will at least convey to your intelligence the fact that *these letters were never meant, and are not now meant, to be other than a quarry of materials from which the book may be drawn?*"[7] It is not surprising that a professional writer such as Stevenson would view his newspaper articles as the "quarry" from which he would build a more enduring structure. But the depth of that conception was unusual. In the main, Stevenson's work went into hard covers with little structural revision on his part; that is, what he published in magazines was either the serial version of his longer fiction, or the essays and short stories that he collected when he had enough to make a volume. In the case of the South Seas, however, he was particularly conscious of the depth and scope of the material, and he was absolutely firm that the newspaper "letters" were merely the first stage of a major work.

I want you to understand about this South Sea Book. The job is immense; I stagger under material. . . . Such a mass of stuff is to be handled if possible without repetition—so much foreign matter to be introduced, if possible with perspicuity—and as much as can be a spirit of narrative to be preserved: You will find that come stronger as I proceed, and get the explanations worked through. Problems of style are (as yet) dirt under my feet: my problem is architectural—creative—to get this stuff jointed and moving.[8]

Stevenson keeps repeating in his correspondence that he is overwhelmed with material, that the problem is bringing it under control and into focus. It is an "architectural" rather than a stylistic problem; indeed, Stevenson observes, "anybody might write of it, and it would be splendid," if only they could organize the prodigious mass of data. Stevenson's unpublished Pacific journal runs to well over two-hundred folio pages, in his very small, even tiny, script. He was intensely conscious of the inherent interest in the incidents and observations he was recording, and he was aware that in order to be successful he would have to devise a method for managing the material that was both reflective and supportive of its nature. Clearly he was working on a scale that the *Sun* and Colvin were incapable of recognizing. They had no appreciation for the nature of the material that Stevenson had at his command. In one sense, it had no meaning to them as matter to be written about except insofar as it possessed an exotic, quaint, or picturesque quality. The kind of serious inquiry that characterized all of Stevenson's work and that he applied to the Pacific adventure was genuinely beyond their understanding. That Stevenson could be interested in island histories, in myths and legends, in social and cultural practices, was hardly conceivable to Colvin, for example, who believed the novelist had sailed off the flat surface of the earth and entered an aboriginal abyss.

The people who were contracting with Stevenson (not to mention his wife, Fanny, and his stepson, Lloyd Osbourne) were simply unable to acknowledge his point of view: what he saw they could not see, and what they could not see (not having been there) they preferred to imagine. And it was an imagining that had little to do with Stevenson's interests. He could never have provided a travelogue of the South Seas, full of exotic descriptions and amusing accounts of foreign lands, the kind of book typified by *Rambles in Polynesia*, published under the name of "Sundowner" in 1897:

> There is an air of happiness about everything in the South Pacific. The palm-trees rustle friendly greetings to the stranger; the birds and animals of the groves and jungles stand their ground as a stranger approaches, confident that no harm is coming to them. The islands themselves are cordial, affectionate, and lovable, honest as the sun, and innocent as doves. Those who have spent time in the Pacific islands grieve at leaving them. Those who have left them are always longing to return. . . . The present denizens of the Polynesian archipelagoes know nothing, and doubtless care less, about where they sprang from. Probably, if you asked them, they would, like the artless Topsy in "Uncle Tom's Cabin," say that they "spected they grow'd." And there is but little hope of solving the mystery, so that it must be allowed to rest. But there they all are in their pristine simplicity, without vice, temper, or waywardness. No women of any colour or kind are more beautiful or tender-hearted than theirs, no men on earth more modest or brave. Christianity has done much to improve and develop their mental faculties but the incursion of the white man, generally speaking, had done them a certain amount of harm. . . .

"Samoan Woman," frontispiece to George Turner's
Samoa a Hundred Years Ago and Long Before (1884).

148

The Polynesian men and women are great, big, lively babies, all mirth and innocence: a curse be on the head of the thoughtless or vicious European aggressor who does anything to stifle that simple mirth or pollute that sacred innocence.[9]

It would be foolish to contrast this with anything in Stevenson; the point is actually more banal. "Sundowner" suitably reflects the expectations of readers confronted with material from a place as foreign and strange as the South Seas. Colvin himself expressed these prejudices: the Polynesian as child, as primitive innocent, not savage because insufficiently vicious to be savage. The picture was absurd even by late Victorian standards, but its publication alone was comment upon its standing. It would be fine to say that nothing in that description connects with reality, but the reality principle can be overrated: who after all would know the difference? Of course the merchant seamen who depended upon the nautical directories and who actually anchored in the islands would know the difference; the sailors on the American, British, and German men-of-war would know the difference; and earlier writers such as Herman Melville and Charles Warren Stoddard knew the difference. But these people could not determine the responses of the readership of the *Sun*. Andrew Lang might well have written to Stevenson and told him that Captain Bosanquet (later Admiral Bosanquet) of the Royal Navy was "fanatical" about his South Seas sketches that were appearing in *Black and White*.[10] But Captain Bosanquet was not *Black and White*'s common reader. Again, there is no question that years after Stevenson's death those who had any knowledge of the Pacific looked to his South Seas writing for authority, confirmation, and/or support. But that class of reader was never in the majority. It must not be forgotten that Stevenson was writing at the end of the nineteenth century, when colonial and imperial expansion was at the center of European politics. Stevenson's offhand remarks (in his initial letter to Colvin) about French colonialism in the Marquesas and British Imperialism in India were hardly profound observations, although they were not popular or common ones. They were the stuff of politics in the 1890s, and for Stevenson, moving into a region as volatile as the Pacific, their reality was immediately apparent.

The letters that Stevenson forwarded from the South Seas and that were printed in the *Sun* constitute a formidable contemporary account of widely scattered archipelagoes and distinctly individual populations and cultures. He wrote of the Marquesas, the Paumotus, the Gilberts, and the "Eight" islands (more commonly known as Hawaii).[11] The range of his discourse, figuratively speaking, was almost as broad as the geographic distances he traversed. Since an exposition of Stevenson's complete text is outside the scope of this essay, I will confine my references to the Marquesas, which

149

were the first group of islands that Stevenson visited. They were the subject of "letters" published between February 1 and April 5, 1891, and they comprise, with the exception of one final "letter"—"The Two Chiefs of Atuona"—the slim volume that McClure printed in 1890 under the title *The South Seas*. Not only does the Marquesan section form the first unit of Stevenson's South Seas discourse, it is representative both in style and substance of the entire work. Issues that Stevenson explores in all the archipelagoes, from the beauty of landscape to the experience of exile, are first introduced here. He commences his ethnohistorical investigation into island cultures, seeking informants who might answer and/or verify his speculations about social and political organization, beliefs and superstitions, education, religion, and folklore. And most importantly, he graphically illustrates the sudden yet systematic decline and death of the Marquesan population, thereby confronting his reader with the consequences of a century's intrusion into the Pacific by white explorers, colonialists, and traders.

Stevenson's section on the Marquesas Islands was broken down into three subheadings: the central island of Nukahiva, with the bays of Anaho, the initial anchorage for the *Casco*, and Hatiheu; next, the village of Tai-o-hae, the French administrative center for the entire archipelago; and finally, the island of Hiva-oa, with the villages of Atuona and Taahauku. In Stevenson's manuscript notes this division is repeatedly reworked, but the principles are consistent through the different versions: Anaho, as the site of the first landfall, receives preliminary treatment; Tai-o-hae as governmental center is the next subject for investigation; and Hiva-oa, with its more remote and perhaps more barbaric traditions, forms the third leg of the triangle. Thus the three subdivisions within the Marquesas grouping are centered on geographic sites, which are then used as architectural masses. This triadic design is a microcosm of the larger book, which is constructed around three separate island chains: Stevenson moves from the Marquesas to the Paumotus to the Gilbert islands, the defining organization of the posthumous *In the South Seas*.

On one level the narrative appears to follow the route of the *Casco*. But Stevenson did not write about Tahiti, where he spent a considerable amount of time; and he absolutely refused to write about Honolulu ("I would rather die"). It is very clear that Stevenson eschewed mere chronology:

How can anybody care when or how I left Honolulu? This is (excuse me) childish. A man of upwards of forty cannot waste his time in communicating matter of that degree of indifference. The letters it appears are tedious: by God, they would be more tedious still if I wasted my time upon such infantile and sucking-bottle details. If ever I put in any such detail, it is because it leads into something or serves as a transition. To tell it for its own sake, never! The mistake is all through that I have told too much; I had not sufficient confidence in the reader, and have overfed him.[12]

In fact, Stevenson completely misread his audience. He himself was an extremely hardworking writer whose exceptional lucidity was a consequence of his unwavering devotion to the craft of revision. Immersed as he was in his work, he lost sight of the complexity of his material, and failed to recognize the degree of attentiveness it demanded of the reader.

"An Island Landfall," the opening chapter of *The South Seas*,[13] is a reflective discourse on the origins of Stevenson's journey. Impelled by illness and the presentiment of death ("For nearly ten years my health had been declining . . . I believed I was come to the afterpiece of life, and had only the nurse and undertaker to expect") Stevenson's departure was a defiance of death and a turning toward life, a rejection of the old world of Europe for the new world of the Pacific. In his narrative, Stevenson dispenses with the obligatory introduction of the crew and passengers; he does not list the points of embarkation and destination, nor does he describe the yacht and the attendant costs of travel. In short, the kind of personal information that Colvin believed most readers would be interested in is altogether absent. Yet Stevenson does introduce an autobiographical mode, which he maintains throughout the narrative. It is a mode that is not strictly speaking "I, Stevenson," but rather that of the stranger as eyewitness to another culture, the European observer who reports back home to his like-minded constituency. Stevenson is well aware of the enormous appeal of the South Seas:

> No part of the world exerts the same attractive power upon the visitor, and the task before me is to communicate to fireside travellers some sense of its seduction, and to describe the life, at sea and ashore, of many hundred thousand persons, some of our own blood and language, all our contemporaries, and yet as remote in thought and habit as Rob Roy or Barbarossa, the Apostles or the Caesars. (*The South Seas* 2)

But even Stevenson's aim, presumably what the *Sun* wanted, is represented in terms that are already distinct from the norm. For one thing, he identifies the people he encounters as "contemporaries" and in the same breath declares them as "remote" as any European ancestors. The effect of this statement is complex. On one level, it does not assume that the people of Polynesia are inferior beings, an attitude that automatically set Stevenson apart from the traditional European traveler. On a second level, the assertion that historical personages from Europe's classical and modern past—from saints and warriors to lawgivers and outlaws—are equally distant in "thought and habit" introduces a subversive element to the discourse. Stevenson disputes the European pretension that its past is one long, unbroken chain that includes Caesar as well as Rob Roy. In effect, he implies that the European past may be as different from the present as its present is from anything that can be found in Polynesia. What Stevenson is questioning is the belief that there is anything so stable or permanent as a great tradition.

That which we call our heritage is nothing more than one particular culture's preserved and protected icons, or idols. The *strangenesses* of other peoples—and *strange* is a word Stevenson uses repeatedly, and not just for Polynesians, for the *Casco*'s crew is equally strange to the islanders—are merely the differences of culture, which are partly determined by biology and geography, and partly by custom and history. Stevenson is conscious of entering a world that never heard the names or events that schoolchildren in Europe grew up with:

> But I was now escaped out of the shadow of the Roman empire, under whose toppling monuments we were all cradled, whose laws and letters are on every hand of us, constraining and preventing. I was now to see what men might be whose fathers had never studied Virgil, had never been conquered by Caesar, and never been ruled by the wisdom of Gaius or Papinian. (*The South Seas* 6)

In his unpublished journal Stevenson writes in the present tense and the note is more immediate—"At last I am to escape out of the shadow of the Roman Empire"—as he relishes leaving the remnant of an old, restrictive civilization behind while anticipating the freedom of a new world.[14]

From the very beginning, Stevenson was aware of the relativism of cultural values, as he wryly observes in a parenthetical aside: "Tropic isles await me, and races truly alien, and civilisations (or, if you prefer the self-righteous word, barbarisms) of a strange descent" (*Journal* 4). Yet despite his conviction that no culture held primacy over another, Stevenson was also conscious of entering a world so different from his own that he wondered how he would ever understand it. "These strange fellow creatures are to pass before me dumb, like visages; or if there be possible any broken talk, it can scarce convey a copious and never a just impression. In seven years, I might acquire some human knowledge; after seven months, I shall have but dipped into a picture book" (*Journal* 4). Stevenson was acutely aware of the role language played not just in primary communication but as an instrument for bridging cultural differences. Unlike previous travelers, he believed that an island culture might be beyond a European's understanding, that it might be impenetrable simply because of linguistic incomprehension. Thus his second chapter, "Making Friends," begins with an extensive discourse on the languages of Polynesia, along with an admission that he had "overestimated" the prospective obstacles.

Stevenson's fluency in French helped him in the Marquesas, and he found that English was far more widely used than he would have expected. Additionally, pidgin English, or "Beach-la-Mar," was rapidly becoming the "tongue of the Pacific." But all this knowledge came after the fact. Before the *Casco*'s first landfall, Stevenson was intensely afraid that he might never comprehend the world he was destined to enter, and for a man whose entire life was devoted to interpretation the thought filled him with regret and a sense of wasted effort.

> I begin to think my money squandered on this cruise: it had been better spent to make me young again, where I might have tasted all the pleasure of discovery in a ten-mile walk. But we are not bound for Bimini, only for Nouka-Hiva and Roua-poa, in whose innocent and stammering names, such as children invent at their play, we may divine the story of a sheltered race, not polished or finished in the contest for existence. (*Journal* 4)

But in spite of his ambivalence about the entire venture, Stevenson reminds himself of the realities: he is not after all on a journey to discover endless youth but merely headed for landfall on some charted islands in the Pacific Ocean. The romantic or imaginative mystery is thus counterpointed by the physical actuality. No matter how personal Stevenson's voice may become, he never loses the ability to step outside himself and place the experience in a larger intellectual context, in this case a context defined by Darwin and the theory of natural selection.

At the start of the journey, and just before the landfall at Nukahiva, Stevenson was clearly more apprehensive than after.

> Not one soul aboard the Casco had set foot upon the islands or knew, except by accident, one word of any of the island tongues; and it was with something perhaps of the same anxious pleasure as thrilled the bosom of discoverers that we drew near these problematic shores. The land heaved up in peaks and rising vales; it fell in cliffs and buttresses; its colour ran through fifty modulations in a scale of pearl and rose and olive; and it was crowned above by opalescent clouds. The suffusion of vague hues deceived the eye; the shadows of clouds were confounded with the articulations of the mountains; and the isle and its unsubstantial canopy rose and shimmered before us like a single mass. (*The South Seas* 3)

The beauty of the island—which is part of the seductiveness of the South Seas and which Stevenson must convey to the reader—is both heightened and qualified by the uncertain and problematic nature of an experience that, like first love, can never be repeated. And nothing about the first South Sea island is quite what it appears to be. The rising of the sun, the color of the trees, the iridescence of the clouds—all are hard to identify precisely. The beauty of the scene is bound up with its insubstantiality, with the inability to predict just what the landscape will look like at the next moment. For the colors change, and the clouds move, and the rains come, and the very sameness of the scene is paradoxically a sign of its transiency.

Stevenson is taken with the idea of ambiguity rather than certainty, with the realization that in attempting to appropriate the islands, as discoverers (or mere travelers) in a long tradition have done, they have in turn been appropriated by them. Thus Krusenstern, whose extended report of the Marquesans appeared in his *Voyage Round the World* (1813), was upbraided by Stevenson for his "innocent" description of what he encountered or saw. And Melville, whom Stevenson considered a genuine writer of genius, was never able to catch the subtleties of island languages. The attempt at appropriation is made difficult, if not impossible, because it demands a reading of the Other, and that reading can never be anything more than speculative. When Stevenson describes the Marquesans as they descend upon the *Casco*, it is predominantly interpretive. "The eyes of all Polynesians are large, luminous, and melting; they are like the eyes of animals and some Italians." But the next sentence was deleted from the newspaper serial: "The Romans knew that look, and had a word for it: *occuli putres*, they said—eyes rancid with expression" (*The South Seas* 6). The deletion removes the note of accusation, not to mention the image of corruption. For a cruise to the South Seas, for the common reader, cannot be presented as a journey to the interior of the soul, but must assume the guise of a voyage to paradise.

Stevenson, however, knew better. "A kind of despair came over me, to sit there helpless under all these staring orbs, and be thus blocked in a corner of my cabin by this speechless crowd; and a kind of rage to think they were beyond the reach of articulate communication, like furred animals, or folk born deaf, or the dwellers of some alien planet" (*The South Seas* 6). This idea holds a powerful sway over Stevenson. He recognized that cruising through the South Seas would not be a simple travel experience. And he was particularly aware that the image of paradise was a problematic one. Although it is now common to view Stevenson's attitude toward the Pacific Islands as a political one—that is, antagonistic to the colonial positions adopted by European governments in the late nineteenth century—it is closer to the truth to see it as deeply philosophical. He was more concerned with the nature of experience, both individual and social, and the problematics of perception, than with realpolitik. As he said of his protagonist Wiltshire in *The Beach of Falesá*, he too was engaged in a "voyage of discovery." In the beginning he believed he would return to England after he recovered his health. But he never went back. Instead he was invigorated by the new possibilities for imaginative freedom, and liberated from the monuments of European history. Somewhere in the course of his journey into self-exile—and that was precisely his understanding of that journey—he realized that he had himself become a "bondslave" to those enchanted isles.

NOTES

1. Henry James to Graham Balfour, November 15, 1901; National Library of Scotland, Balfour Papers, MSS. 9895ff.192–95.
2. Hubert Howe Bancroft, *The Native Races*, 5 vols. (New York: 1874–76).
3. See Roger Swearingen's discussion of the composition of "The South Seas" in *The Prose Writings of Robert Louis Stevenson* (Hamden, CT: 1980), 139–43.
4. S.S. McClure to R.L.S.; Yale University, Beinecke Library. Subsequent references to materials from the Beinecke Library will appear within the text as Beinecke.
5. This rare book was published in early November 1890 in an edition of twenty or twenty-two copies primarily for copyright purposes. Swearingen dates the publication November 12 (140), although a letter in one of the Beinecke copies from Cassell and Company (May 4, 1939) cites November 2 as the date. A transcription error could easily explain the discrepancy.
6. Undated letter [July 1888], Harvard University, Houghton Library. In the published version Colvin substituted "even" for "raw" and "strangeness" for "strangenesses" (*The Letters of Robert Louis Stevenson*, ed. Sidney Colvin, Vailima Edition, vol. 22 [New York: 1923], 101).
7. RLS to SC, May 17 [1891]; Houghton.
8. May 17 [1891]; Houghton.
9. London: European Mail, Limited, viii, 5–6, 8.
10. Andrew Lang to RLS, September 21 [1891]; Beinecke.
11. For the latter he concentrated on the big island of Hawaii and the small island of Molokai, the site of the leper colony at Kalaupapa. *In the South Seas*, however, published two years after his death in 1894, dropped the Hawaiian material and all subsequent editions sustained that excision. Thus Stevenson's moving description of his visit to Kalaupapa—including his refusal to put on gloves as a protective device when he shook hands with the lepers—can only be found in specialized texts that reproduce the original newspaper letters.
12. RLS to Sidney Colvin, Tuesday, Houghton.
13. London: 1890. Subsequent references to this volume will appear parenthetically within the text as *The South Seas*. The newspaper letters (*Letters*) had no chapter titles, only Roman numerals indicating the divisions.

14. Stevenson's unpublished Pacific journal (*Journal*) is in the Huntington Library, bound in two large volumes. Subsequent references will be parenthetical within the text.

AFRICAN LANDSCAPE AND IMPERIAL VERTIGO

David Carroll, University of Lancaster

In 1911 the proconsul Sir Harry Johnston published a volume in the Home University Library entitled *The Opening Up of Africa*. It is a short study that seeks to describe from prehistoric times to the present the crucial events in African history, namely the sequence of European penetrations:

> The white man in his Mediterranean and Nordic types has been attempting to penetrate Africa for many thousands of years, for ten thousand years, at any rate; and it is this story of his extraordinary perseverance in a dire struggle with great natural forces that I propose to tell in these pages. (10)

It is, however, a story that is curiously foreshortened because on the first page of his history Johnston states unequivocally that twenty-five years ago Africa—"a region of the earth's land surface which the explorer Stanley had just nicknamed 'the Dark Continent'" was virtually unknown in Europe and America. What has prompted his present study is the recent extraordinary transformation:

> since 1885 African discovery has proceeded at a rate so astonishing that there is nothing quite comparable to it in the history of human civilization. The white man has been aided by his own wonderful inventions to traverse the wilderness and not die, to combat hostile human tribes—some of them separated from him in culture by fifty thousand years—to

exterminate wild beasts and brave the attacks of far more deadly insects and microbes. (9)

This recent period is the culmination of Europe's longstanding attempt to rescue the "great mass of the Negro sub-species . . . locked up in the African Continent south of the Desert Belt" from its long suicide of isolation in which it "stagnated in utter savagery." Indeed, thousands of years ago, Africans "may even have been drifting away from the human standard back towards the brute when migratory impulses drew the Caucasian, the world's redeemer, to enter Tropical Africa by four main routes" (24–25).

Coming from a man with extensive Imperial experience in virtually all areas of Africa and whose crowning achievement, as recorded in the *Dictionary of National Biography*, is that he "was actively instrumental in adding about 400,000 square miles of the African continent to the British Empire," such views have a vivid, exemplary quality. I quote them to draw attention to some of the historical, cultural, and racial assumptions that motivated the scramble for Africa that was formalized at the Berlin Conference in 1885. Africa is separate; it is prehistoric; it is dangerous; but, despite itself, it will be opened up and thereby redeemed by Europe. These attitudes have, of course, been studied in great detail and perhaps most revealingly by Philip Curtin in *The Image of Africa* (1965).[1] Needless to say, not all writers shared the proconsul's confidence. The limited aim of this essay is not to reexamine such attitudes, but to see how they are reflected and questioned in some English fiction at the turn of the century. The focus of this discussion will concern a crucial point of contact between the European explorer or colonial official and the Dark Continent—the African landscape—and the following question will be addressed: in depicting these strange locations, how did the writers represent the alien and seek to make it intelligible? The strategies are various, but all of them tell us a good deal about English values, English assumptions, and English fiction during a crucial period of its development.

It is worth pointing out that by the last quarter of the century the heroic age of African exploration was over. The first extensive European penetration into the interior of the continent had been carried out by such figures as Heinrich Barth, David Livingstone, Richard Burton, John Hanning Speke, and Samuel White Baker, whose accounts of their adventures had been hugely popular. Now in the 1870s and 1880s the commercial "opening up" of the continent was beginning, made possible by two key technological developments: river steamers had overcome the obstacle of poor transportation, and quinine could control the deadly effects of the mysterious malaria. Later, during the phase of conquest, the rapid-firing rifle and the machine-gun were added as the crucial tools of empire.[2] As a result, the scramble that followed the Berlin Conference rapidly transformed the map of Africa and fixed the political boundaries that are still in force today. In its train came the writers—scientists, explorers, novelists, anthropologists—keen to capitalize on the growing popularity of Africa as a subject of both scholarly and melodramatic interest.

At the end of the century, which predecessors could these writers, keen to respond to the popular interest in Africa, learn from? There were the explorers' accounts such as Livingstone's *Missionary Travels* (1857), the several narratives of the various Niger expeditions earlier in the century, and H.M. Stanley's more melodramatic *How I Found Livingstone* (1872) and *In Darkest Africa* (1890). In addition, there was the boys' adventure tradition initiated by G.A. Henty with his novels describing the Ashanti campaigns of 1874, *By Sheer Pluck* (1884) and *Through Three Campaigns* (1903). But there was one major English novelist who in the 1870s was already depicting remote communities living amid strange landscapes and threatened by unwelcome interlopers. This was, of course, Thomas Hardy, and his novels have a direct relevance to this subject.

First of all, Hardy radically questioned the adequacy of classical pastoral conventions to present primitive landscape in the modern world, most memorably in the account of Egdon Heath in the opening pages of *The Return of the Native* (1878). Described in terms of twilight anomalies, Egdon fits into no conventions: it is both day and night, "a division in time no less than a division in matter" (3), both Genesis and Revelation, alien but attractive. The heath has always been there but it is only now coming into its own: "Haggard Egdon appealed to a subtler and scarcer instinct, to a more recently learnt emotion, than that which responds to the sort of beauty called charming and fair" (5). The instinct and emotion are those of the apocalypse for, after all these centuries, Egdon "could only be imagined to await one last crisis—the final overthrow" (4). The heath appeals to a realization which Hardy is seeking to place precisely: "Indeed, it is a question if the exclusive reign of this orthodox beauty is not approaching its last quarter. The new Vale of Tempe may be a gaunt waste in Thule: human souls may find themselves in closer and closer harmony with external things wearing a sombreness distasteful to our race when it was young" (5). In this end time modern consciousness requires new landscapes to mirror its alienation. It is a well-known passage, but it takes on a fresh significance in this context. Hardy is describing a prehistoric landscape that cannot be contained within literary conventions and in doing so rehearses brilliantly all the strategies lesser writers will deploy in more distant locations. Finally, they too will turn, like he does, from external description to the darkness within to which the landscape appeals: "Then it became the home of strange phantoms; and it was found to be the hitherto unrecognized original of those wild regions of obscurity which are vaguely felt to be compassing us

159

about in midnight dreams of flight and disaster, and are never thought of after the dream till revived by scenes like this" (6).

Second, Hardy makes a clear distinction between the natives and the interlopers: the heath only disturbs the latter. "Civilization was its enemy" (6–7) and it is those with a civilizing mission who learn to fear it, recognizing "that sinister condition which made Caesar anxious every year to get clear of its glooms before the autumnal equinox" (62). The natives, on the other hand, phlegmatically light their bonfires and carry out their rituals with the boredom which, Hardy says, is the mark of genuine custom. It is this extreme contrast between the insiders to a culture and a landscape and the outsiders which makes Hardy the most anthropological of Victorian novelists. His narrator is the bridge between the sophisticated reader and the half-fictional, half-real world of Wessex carefully mapped in the novels; and one of his anthropological tasks is to put history, memory, and meaning back into the landscape. The difficulties of achieving this are articulated through what are essentially three groups of characters. There are the modern sophisticates and romantics, the newcomers like Dr Fitzpiers in *The Woodlanders* (1887), who, unable to understand the world of the woodlands, quickly comes to feel abandoned: "the loneliness of Hintock life was beginning to tell upon his impressionable nature" (155). Then there are the natives who inhabit a world of custom and memory. What they draw upon which he lacks are "old associations—an almost exhaustive biographical or historical acquaintance with every object, animate and inanimate, within the observer's horizon" (155). This enables them to fill their locale with significance. But there is also a more secret knowledge only to be learned deeper in the woodlands. This is not the symbolism of human association but the secret language of the laws of nature. Giles and Marty "had been possessed of [the woodland's] finer mysteries as of commonplace knowledge; had been able to read its heiroglyphs as ordinary writing . . . together they had, with the run of the years, mentally collected those remoter signs and symbols which seen in few were of runic obscurity, but all together made an alphabet" (415). Their secret is also their weakness when the interlopers appear on the scene. Finally, there are those central figures in Hardy's fiction who seek to negotiate between the primitive and the modern, attempting to merge the horizons of these two separate worlds. This is always a dangerous game. Sometimes, like Grace Melbury, they survive with a newfound pragmatism; more often, like Tess and Jude, this hybrid status and its oscillations prove fatal.

This simplified model will enable us to examine some of the accounts of African, and especially West African, landscape described and fictionalized in the period from 1876 to 1914. From what point of view is it described? That of the interloper, the native, or the intermediary? Implicit in this model are the stages of a journey on which the narrator conducts his middle-class readers: first they are separated from their normal life in their own culture, next they enter a limbo or liminal world outside the known social system, and then they are re-aggregated into their own culture. These stages are analogous to those defined by Arnold van Gennep and developed by Victor Turner in their analysis of the transition rites which accompany every change of state, social position or age.[3] The crucial stage is the betwixt-and-between stage of liminality, which is seen as outside any social system, a timeless condition where all values are questioned and things become their opposites. In this phase, individuals, freed from social convention, catch a glimpse of their true place in the cosmos.

Such an idea of the rite of passage is integral to much European writing about Africa. And it is not only the explorers or travel writers who voyage out, penetrate the continent, and exist in the bush before returning home to normality and security; but also the youthful adventurers of popular fiction, the embattled District Officer narrating his memoirs, or the retired missionary. One way of assessing these writers is to see to what extent they are aware of the implications of this basic model; and in particular to what extent are these implications brought out in their descriptions of the African landscape. To give some idea of the variety of mode and response during the period, two works of popular fiction will be briefly examined, along with an account by a famous travel writer, and, finally, Joseph Conrad's two fictions set in Africa—a short story and the novella that has acquired canonical status in any discussion of Europe's Africa.

Popular fiction of the period, representing as it does the unquestioning, extrovert face of Imperialism, is not too concerned with the complexities and ambiguities of the journey into the unknown. The hero, often boyish, sets off for Africa on the apprenticeship of life, penetrates into forbidden and dangerous places before returning with his treasure or reputation to the civilized worlds. If Hardy represents some of the growing uncertainties and self-questionings of European culture, these writers provide the simple myths of British Imperial aggression. Yet even here the structure of the myth demands its liminal phase, that area of uncertainty in which normality is put at risk, and this is again often depicted through landscape, though it is landscape primarily as the scene of adventure. Two such novels, which span the period of intense imperial expansion described by Sir Harry Johnston, are Rider Haggard's *King Solomon's Mines* (1885) and John Buchan's *Prester John* (1910). Both are set in southern Africa and both describe journeys into an alien landscape where African kingdoms have been or are in the process of being established. In the former novel the usurpers have to be replaced by the rightful king and the kingdom of death overthrown; in the latter the black Christian minister and heir to the mythical Prester John, Laputa, has to be prevented from establishing his own kingdom.

The search for King Solomon's mines leads the expedition across the desert and through the mountains, "Sheba's snowy breasts," to that seductive Edenic panorama beloved of European explorers: "The landscape lay before us like a map, wherein rivers flashed like silver snakes, and Alpine peaks crowned with wildly twisted snow wreaths rose in grandeur, whilst over all was the glad sunlight and the breath of Nature's happy life" (111). But appearances are deceptive. Kukuanaland is under the usurping rule of Twala and his deadly prophetess, Gagool. It has become the kingdom of death that the explorers have to penetrate and destroy, passing the three vast statues of the strange gods after which Solomon went astray, and entering the cave where all previous kings, transformed into stalactites, are presided by death itself, "shaped in the form of a colossal human skeleton, fifteen feet or more in height" (294). Beyond, in the heart of the kingdom, is the treasure chamber where the Europeans are finally trapped in darkness, buried alive with the gold and diamonds. Momentary despair is overcome briskly with European logic; they extricate themselves and hand over the kingdom to its rightful and enlightened ruler. They return home confirmed in their own values having crossed several boundaries, including the chamber of death at the heart of the continent. There is only a brief suggestion of a fatal attraction. One of the explorers falls in love with a royal beauty who is inadvertently killed during their adventures, but the liaison had little future for as she herself put it, "Can the sun mate with the darkness, or the white with the black?" (333).

In *Prester John* the penetration of the African landscape also culminates in a cave, but here the situation is more complicated. David Crawfurd, witnessing the anointing and crowning of Laputa with the necklet of Prester John, becomes spellbound by the ex-minister's syncretist prayers compounded of biblical texts and black nationalism: "I heard the phrases familiar to me in my schooldays at Kirkcaple. He had some of the tones of my father's voice, and when I shut my eyes I could have believed myself a child again" (142). An earlier self is being reshaped and transformed in a new way and he is overwhelmed by the experience:

> My mind was mesmerized by this amazing man. I could not refrain from shouting with the rest. Indeed I was a convert, if there can be conversion when the emotions are dominant and there is no assent from the brain. I had a mad desire to be of Laputa's party. Or rather, I longed for a leader who should master me and make my soul his own, as this man mastered his followers. (145)

But the white spy is discovered and, prevented from going native, he sets

"Up above them towered his beautiful pale face," Charles Kerr, illustration to
Rider Haggard's *She* (1913).

about defeating Laputa and his rebellion: "Last night I had looked into the heart of darkness, and the sight had terrified me" (151). As is customary in fiction of this kind, the successive stages of the hero's apprenticeship are symbolized by the stereotyped minor characters: the timid schoolmaster repelled by Africa's menace, the sleazy Portuguese who has gone over to Laputa, and representing the class of double agents, the military man who can quickly transform himself with black coloring into a native and back again. All these represent the relatively fixed alternatives through which the hero progresses.

The climax of the adventure occurs in the same cave from which, after Laputa's death amid his treasure, the hero has to extricate himself in order to warn his compatriots of their danger. After a superhuman climb he escapes only to experience again the ambiguities of the primitive, but this time in the form of landscape.

> In three strides I was on the edge of the plateau. Then I began to run, and at the same time to lose the power of running. I cast one look behind me, and saw a deep cleft of darkness out of which I had climbed. Down in the cave it had seemed light enough, but in the clear sunshine of the top the gorge looked a very pit of shade. For the first and last time in my life I had vertigo. Fear of falling back, and a mad craze to do it, made me acutely sick. I managed to stumble a few steps forward on the mountain turf, and then flung myself on my face. (264)

The moment passes, the great rebellion is defeated, and Crawfurd returns home aware of the white man's burden: "That is the difference between white and black, the gift of responsibility, the power of being in a little way a king." And he now knows how that responsibility should be exercised having seen "something of [the Africans'] strange, twisted reasoning" (276). Someone has to undertake the risk and the twists, then overcome the subsequent vertigo—that moment of attraction and repulsion—for this secret knowledge to be used for everyone's benefit.

In contrast to the relative simplicities of the adventure story are the works of Mary Kingsley, a writer who explores the uncertainties of the African landscape and its implications. Like Hardy she is fully aware at the end of the century of the ambiguous position of the interloper seeking to decipher alien codes. Her *Travels in West Africa* (1897) questions and parodies the conventions of African travel and in doing so brings to the genre a more sophisticated awareness of the European writer's role. As she describes the forests of the French Congo, for example, she enacts all stages of the journey through her multiple points of view:

164

To my taste there is nothing so fascinating as spending the night out in an African forest, or plantation; but I beg you to note I do not advise any one to follow the practice. Nor indeed do I recommend African forest life to any one. Unless you are interested in it and fall under its charm, it is the most awful life in death imaginable. It is like being shut up in a library whose books you cannot read, all the while tormented, terrified, and bored. And if you fall under its spell, it takes all the colour out of other kinds of living. (102)

You can keep out and retain your European sanity; or you can enter the forest without understanding its language and begin to lose your bearings ("tormented, terrified, and bored"); or you can fall under its spell, cross the threshold, as it were, and learn the language of the forest, but the cost is high. Then your other life is drained of meaning. It is the same with the natives. If you live among them "you gradually get a light into the true state of their mind-forest. At first you see nothing but a confused stupidity and crime; but when you get to see—well! as in the other forest,—you see things worth seeing" (103). As in Hardy, the one person who is able to negotiate the complete return journey is the writer who can deploy vivid reportage, empathy, irony, whimsicality, and a parodic questioning of the conventions of travel narrative, in order to cross the threshold, briefly be assimilated into the alien world, and then disengage in order to return home.

All these skills are evident in her memorable chapter "Voyage Down Coast": "Wherein the voyager before leaving the Rivers discourses on dangers, to which is added some account of the Mangrove swamps and the creatures that abide therein, including the devil of an uncle." The centerpiece is a lengthy account of the great mangrove swamps of West Africa, which brings out vividly the paradoxes, ambiguities, and contradictions of the African landscape. First of all, a mangrove swamp at the mouth of a great African river was invariably the border territory marking off one stage of the journey (the sea voyage) from the next (penetration up river). As a threshold it is a vivid example of something that falls between the accepted categories, in this case of land and sea, the boundaries of which are constantly shifting as the mangroves colonize the estuaries and then die. It is a landscape of mud in which walking and canoeing are equally dangerous. Other European conventions are ignored in this strange, liminal terrain. She quotes approvingly Lord Lugard's depiction of the monotonous, green-black walls of mangroves "as if they had lost all count of the vegetable proprieties, and were standing on stilts with their branches tucked up out of the wet, leaving their gaunt roots exposed in mid-air" (87). As with the Congo forest this is a landscape not easy to read and yet it commandeers all other realities in its surreal ambience:

I shall never forget one moonlight night I spent in a mangrove-swamp. I was not lost, but we had gone away into the swamp from the main river. . . . We got well in, on to a long pool or lagoon; and dozed off and woke, and saw the same scene around us twenty times in the night, which thereby grew into an aeon, until I dreamily felt I had somehow got into a world that was all like this, and always had been, and was always going to be so. (92)

As with Egdon Heath, the moonlight or twilight landscape becomes a spatial representation of timelessness, which erodes all differences and severs all previous connections. But parody keeps any Conradian heavy breathings at bay. When a crocodile, or "a mighty Silurian, as *The Daily Telegraph* would call him" gets his front paws over the stern of the canoe, Kingsley briskly discourages it with "a clip on the snout with a paddle" (89).

Like Hardy, she is seeking to capture a new kind of emotion that can only be articulated through the various contraries embodied in these primal scenes. She focuses eventually on one area of the West Coast "much neglected by English explorers":

I believe the great swamp of the Bight of Biafra is the greatest in the world, and that in its immensity and gloom it has a grandeur equal to the Himalayas. I am not saying a beauty; I own I see a great beauty in it sometimes, but it is evidently not of a popular type, for I can never persuade my companions down in the Rivers to recognise it; still it produces an emotion in the stoutest-hearted among them; yea, even in those who have sailed the world round. (95)

Then the focus narrows further to a particular place at a particular time: the Bonny estuary of the Niger on a Sunday in the wet season. Seaward there is the foam of the dangerous bar, everywhere else the endless walls of mangrove in the rotting mud water, only broken by the black ribs of the old hulks once used as trading stations. On her first visit "a sense of horror seized on me as I gazed upon the scene" but the interloper can eventually cross the threshold and embrace the unintelligible: "Five times have I been now in Bonny River and I like it" (97).

One essential feature of Bonny is finally selected for comment, its smell: "That's the breath of the malarial mud , laden with fever, and the chances are you will be down to-morrow" (97). It is difficult to exaggerate the significance of the so-called miasma in any account of Europe's contact with Africa before the link between the Anopheles mosquito and malaria was established in 1897. The explorer penetrates Africa; this mysterious emanation of Africa invades and debilitates the explorer. Mary Kingsley goes on to describe its approach in this way:

> If it is near evening time now, you can watch it becoming incarnate, creeping and crawling and gliding out from the side creeks and between the mangrove-roots, laying itself upon the river, stretching and rolling in a kind of grim play, and finally crawling up the side of the ship to come on board and leave its cloak of moisture that grows green mildew in a few hours over all. (97)

The miasma is the correlative of the alien forces that can take over the self, all the more powerful for being inexplicable and, to Kingsley, attractive. The mist's shifting transformations are so seductive that she has "often, when no one has been near to form opinions of my frivolity, played with it, scooping it up in my hands and letting it fall again, or swished it about with a branch" (418). And yet she devotes an appendix to the diseases of West Africa in which she acknowledges that—despite the effectiveness of the quinine prophylaxis—no one knows the cause of malaria. The so-called experts "tell you, truly enough no doubt, that the malaria is in the air, in the exhalations from the ground, which are greatest about sunrise and sunset, and in the drinking water, and that you must avoid chill, excessive mental and bodily exertion, that you must never get anxious, or excited, or lose your temper" (684). Its deadly effects are baffling; some Europeans have "an unknown element in their constitutions that renders them immune" (690), the rest don't. It exists on the borders between a contagious disease and a mental illness; and Kingsley ends her account not with the dangers of the savages "but with a foe you can only incarnate in the dreams of your delirium, which runs as a poison in burning veins and aching brain—the dread West Coast fever" (691). Here, as it were in the landscape itself, is articulated the other dreaded yet half-longed-for possibility of the journey: that of remaining bewitched in a liminal world of no-place and no-time. Beyond the dangers of going native is the no-return of death.

In Conrad's two pieces of African fiction, contemporaneous with Mary Kingsley's travels, the representatives of European civilization fail to return. In each case the landscape is both the expression and the agent of their destruction. In "An Outpost of Progress" (1897) the two white men, having been cynically abandoned at their trading station up the Congo, pass through all the stages of separation, alienation, and liminality, but without the re-aggregation inevitable in the adventure story. First, there is the loneliness and what it brings: "to the negation of the habitual, which is safe,

167

there is added the affirmation of the unusual, which is dangerous" (89). The landscape embodies this erosion of security and meaning. "They lived like blind men in a large room, aware only of what came in contact with them (and of that only imperfectly), but unable to see the general aspect of things. The river, the forest, all the great land throbbing with life, were like a great emptiness." The river upon which they depend "seemed to come from nowhere and flow nowhither. It flowed through a void" (92). The void is internalized when, without much resistance, they become involved in the slave trade. Now in their culpability they are even more vulnerable; they experience "an inarticulate feeling that something from within them was gone, something that worked for their safety and had kept the wilderness from interfering with their hearts" (107). The contagion has begun and the wilderness is able "to approach them nearer, to draw them gently, to look upon them, to envelop them with a solicitude irresistible, familiar, and disgusting" (29).

The climax comes when they turn against each other after a trivial disagreement and act out a deadly game of hide-and-seek around their verandah. Conrad has turned the great imperial game into a black comedy of the absurd. The two men circle faster and faster until Kayerts experiences the characteristic vertigo before accidentally shooting his colleague:

> Then he also began to run laboriously on his swollen legs. He ran as quickly as he could, grasping the revolver, and unable yet to understand what was happening to him. He saw in succession Makola's house, the store, the river, the ravine, and the low bushes; and he saw all those things again as he ran for the second time round the house. Then again they flashed past him. (111)

The meaningless list of the ingredients that make up their world—the culmination of Conrad's many topographical catalogues—are whirled into the vertigo in which all differences are cancelled. After the shot he doesn't know whether he has killed or is killed. Then, sitting by the corpse, he ponders with lunatic lucidity "the conviction that life had no more secrets for him: neither had death!" This is the limbo in which the self floats free: "He seemed to have broken loose from himself altogether" (114). He falls asleep, wakes in the dawn and is finally claimed by the miasma: "He stood up. The day had come: the mist penetrating, enveloping, and silent; the morning mist of tropical lands; the mist that clings and kills; the mist white and deadly, immaculate and poisonous" (115). His shriek of horror mingles with the hooting of the approaching steamer ("Progress was calling to Kayerts from the river") before he hangs himself.

In *Heart of Darkness* (1902) the stages of disintegration are much more

168

elaborately plotted as we are conducted from a defamiliarized Thames and Brussels, to the West African coast and the Company Station, then on a two-hundred-mile tramp to the Central Station, and finally up river to the Inner Station and Kurtz. Marlow repeats the journey and provides the disorienting commentary on the landscape, each phase of which furnishes a set piece of the pastoral of alienation. And each phase is a threshold to further mystery. On the Thames, for example, we are invited to imagine the feelings of the Roman legionary threatened even there by the dreaded miasma: "Imagine him here—the very end of the world, a sea the colour of lead, a sky the colour of smoke. . . . Sandbanks, marshes, forests, savages . . . cold, fog, tempests, disease, exile, and death—death skulking in the air, in the water, in the bush" (49). Then comes the steamer journey down the African coast with Marlow carefully registering the scenes and his emotions. "Watching the coast as it slips by the ship is like thinking about an enigma" (60). But here on the periphery the enigma remains incomprehensible even though the French man-of-war tries to pound some sense into the continent: "in and out of rivers, streams of death in life, whose banks were rotting into mud, whose waters, thickened into slime, invaded the contorted mangroves, that seemed to writhe at us in the extremity of an impotent despair" (62). After the Company Station, with its grove of death, Marlow's journey becomes a march through the bush, which is equally enigmatic: "Paths, paths, everywhere; a stamped-in network of paths spreading over the empty land . . . and a solitude, a solitude, nobody, not a hut" (70). It is the paradoxes, the absurdity that Marlow stresses as he tracks Kurtz, but on the final stage—the river journey to the Inner Station—there is a significant change. The descriptive set piece ("Going up that river was like travelling back to the earliest beginnings of the world") still consists of paradox: this is a desert in which vegetation riots, a gloom in which the brilliant sun burns down, a stillness noisy with threats (92–93). But then there comes a change, a glimpse of "a whirl of black limbs, a mass of hands clapping, of feet stamping, of bodies swaying, of eyes rolling" and with it the terrible possibility: "Well, you know, that was the worst of it—this suspicion of their not being inhuman" (96). Comprehensibility now becomes the greater horror, the terror "of there being a meaning in [the noise] which you—you so remote from the night of first ages—could comprehend" (96). It is, however, not Africa which is becoming intelligible, but rather the inky depths of the European soul. Stage by stage the landscape provides Marlow retrospectively with the key to Kurtz's destiny and, as it does so, its lineaments become increasingly obscured.

As the rescuers approach the Inner Station, the Congo mist isolates the steamer and they are cut off in a void: "The rest of the world was nowhere, as far as our eyes and ears were concerned" (102). Vertigo is imminent:

"Were we to let go our hold of the bottom, we would be absolutely in the air—in space" (106). But Marlow's moorings *are* slipping and with that comes an awareness of the wilderness itself as the protagonist that has requisitioned Kurtz for its own purposes. "The wilderness had patted him on the head . . . it had caressed him, and—lo!—he had withered; it had taken him, loved him, embraced him, got into his veins, consumed his flesh, and sealed his soul to its own by the inconceivable ceremonies of some devilish initiation" (115). It can do this because "he was hollow at the core." In response to Kurtz's "fantastic invasion," the wilderness has invaded him in turn with its insinuating miasma. The Africans—feet stamping, bodies swaying, eyes rolling—are simply a momentary congealment of the wilderness.

The question in all such adventure stories, even in this parodic version, is: can the white man be plucked to safety from the hostile natives? Or, in slightly different terms, can he move through the liminal stage of danger and anarchy back into the security of the known? It is a movement that is repeatedly attempted and frustrated in this story. Earlier we have had a glimpse of Kurtz setting off for the Central Station with his ivory and then turning back. Now he is carried on to the steamer but escapes back to the jungle in the middle of the night, thereby breaking all the codes of European rescue. The climactic and most dangerous episode—the third attempt to return Kurtz to civilization—characteristically takes the form of a bizarre game. Marlow circles around the fleeing Kurtz in the bush ("I verily believe chuckling to myself") to block his escape: "I was circumventing Kurtz as though it had been a boyish game" (142). Here, with the two men gambolling in limbo, the ultimate moment of truth becomes topsy-turvily a moment of farce, for there is no standard by which it can be measured. How does one rescue someone who has gone completely native? "I tried to break the spell—the heavy, mute spell of the wilderness—that seemed to draw him to its pitiless breast" (144). But Marlow has nothing to which he can appeal. As he had speculated earlier, "The thing was to know what he belonged to, how many powers of darkness claimed him for their own" (115). Now he knows that, on the contrary, everything belongs to Kurtz. "I had, even like the niggers, to invoke him—himself—his own exalted and incredible degradation. There was nothing either above or below him, and I knew it" (144). The wilderness has claimed him but he has finally claimed the wilderness. And then the landscape disappears: "He had kicked himself loose of the earth. Confound the man! he had kicked the very earth to pieces. He was alone, and I before him did not know whether I stood on the ground or floated in the air" (144). Everything is swept up into the vertigo of Kurtz's madness within which Marlow watches "the inconceivable mystery of a soul that knew no restraint, no faith, and no fear, yet struggling blindly with itself" (145).

This struggle is for Conrad what the seductive and destructive African landscape finally symbolizes. The European soul finds there simultaneously its death-wish and its self-assertion. As Kurtz dies, "both the diabolic love and the unearthly hate of the mysteries it had penetrated fought for the possession of that soul" (147). All the stages of separation, alienation, and assimilation have been leading—as hinted at in all these stories—to that final threshold of the unknown of which these others are preparatory. And the threshold only becomes a threshold if one steps over it. Not for Kurtz is there the return with the diamonds, the ivory, or useful knowledge of the natives, but death and the judgment on his own transgression, which he conflates in his final exclamation. Even Marlow, the faithful commentator and secret sharer, here proves inadequate as he acknowledges that crucial distinction between them:

True, he had made that last stride, he had stepped over the edge, while I had been permitted to draw back my hesitating foot. And perhaps in this is the whole difference; perhaps all the wisdom, and all truth, and all sincerity, are just compressed into that inappreciable moment of time in which we step over the threshold of the invisible. (151)

All the other thresholds between the known and the unknown—Europe and Africa, white and black, civilized and primitive—are shown finally to be surrogates for this ultimately liminal moment, out of time and space, for which they have been preparing the way. With deep, late-Victorian irony Conrad locates his bleak apocalypse at the very cutting-edge of the Imperial enterprise as the African coastline, mangroves, and rivers that have beckoned the explorers and traders to their self-destruction disappear into Kurtz's monstrous egoism. The representative of European values is allowed to deliver the final judgment.

NOTES

Quotations are from the following editions: John Buchan, *Prester John* (London: n.d.); Joseph Conrad, *Youth, Heart of Darkness, The End of the Tether* (London: 1902); Rider Haggard, *King Solomon's Mines* (London: n.d.); Thomas Hardy, *The Return of the Native* (London: 1909), *The Woodlanders* (London: 1951); Sir Harry Johnston, *The Opening up of Africa* (London: 1911); Mary Kingsley, *Travels in West Africa* (London: 1897).

1. Philip D. Curtin, *The Image of Africa: British Ideas and Action, 1780–1850* (London: Macmillan, 1965).

2. See Daniel R. Headrick's accessible account of these crucial developments: *The Tools of Empire: Technology and European Imperialism in the Nineteenth Century* (Oxford: 1981).
3. Turner has developed Arnold van Gennep's basic model (described in *Rites of Passage*, 1909) in a variety of anthropological works. See especially his *Dramas, Fields, and Metaphors* (Ithaca, NY: 1974).

ABROAD AS METAPHOR: CONRAD'S IMAGINATIVE TRANSFORMATION OF PLACE

Daniel Schwarz, Cornell University

I

Abroad for Conrad, of course, was England, the country that provided a home and language with which to write. Conrad was in exile, an expatriate from his native Poland, and during the formative years of his adulthood he was a wanderer. Yet it is not an exaggeration to say that his most important journeys took place in his own mind. Some of his most graphically depicted places—Costaguana, the imagined country of *Nostromo* (1904), and the cosmopolitan Russia of *Under Western Eyes* (1911)—are places he never visited, but Russia, which he regarded as the dreaded enemy to Poland's independence, haunted his imagination. That a place could be a source of alienation, marginality, and exclusion derived from his having been born in a section of Poland under Russian rule, and having, in his early childhood, accompanied his father and mother into an exile imposed because of his father's radical political activity in Warsaw. Not until late in his life, if ever, did Conrad quite think of himself as settled in an appropriate place, and that

173

THE ENDS OF THE EARTH: 1876–1918

sense of himself as wanderer, as not belonging even in England, preyed upon his psyche.

I want to use as my point of departure Wallace Stevens's poem, "Mrs. Alfred Uruguay," a poem in which he imagines an elegantly dressed woman undergoing a traditional quest up a mountain which allegorically represents the real. As she makes her way up the mountain on a donkey without companions, she is passed by a lone, poorly dressed man on a horse descending hurriedly from the mountain of the real to the imagined land where experience is reshaped:

> Who was it passed her there on a horse all will,
> What figure of capable imagination?
> Whose horse clattered on the road on which she rose.
>
> . . . .
>
> The villages slept as the capable man went down,
> Time swished on the village clocks and dreams were alive,
> The enormous gongs gave edges to their sounds,
> As the rider, no chevalere and poorly dressed,
> Impatient of the bells and midnight forms,
> Rode over the picket rocks, rode down the road,
> And, capable, created in his mind,
> Eventual victor, out of the martyrs' bones,
> The ultimate elegance: the imagined land.[1]

Is not the man of capable imagination an apt metaphor for Conrad, the man in exile, who negotiates between the reality of his own life and his powerful imagination?

This essay will define how, in Conrad, setting negotiates between anterior reality in which the author lived and found the ingredients for his experience and the imagined world that results from what Stevens called the capable imagination. The process of creating setting, or setting-making, may be thought of as a verb to show how the writer moves between the two poles of anterior reality and the fictive world. Let us consider the nature of setting. Setting is the physical place in which the action and plot of an imagined world takes place. But is it not also the imaginative place to which the desires, hopes, and plans of writers struggle with their doubts, anxieties, and frustrations? Setting is a seam where author and reader meet. Setting not only reflects the action, character, thematic issues, and linguistic patterns of a novel, but also takes its definition from them. In current terms, setting is constituted and constituting; it reflects not only the historical conditions of the imagined world, but also those of the real world in which the author has lived. It is one of the more determinate codes that authors create for readers, and yet, no matter how precise and graphic the description, setting always leaves cracks and crevices for the reader's imagination. Mikhail Bakhtin's concept of chronotype reminds us of the inseparability of space and time: "Time, as it were, thickens, takes on flesh, becomes artistically visible; likewise, space becomes charged and responsive to the movements of time, plot, and history."[2] Novels rescue time from oblivion by giving it the shape of space; they enable us to realize events in tangible readings. The reader recalls setting as a *paysage moralisé*, a mindscape engraved with the events that transpired.

While no amount of precision and nominalism can exclude a reader from creating, as Stevens does when looking at the setting of Key West, ghostlier demarcations and keener sounds, the precision and detail of setting are means by which the author controls and limits the reader. Yet the reader participates in limiting, defining, and clarifying space; as he reads, he goes "abroad" from the text. Setting is perhaps the most likely element to reflect anterior reality because of the authors' investment in the places of their lives, in their narratives of their memories. For Conrad, setting often played a crucial Oedipal role, for it represented the visionary gleam that he was forever deprived of. As we shall see, Conrad's settings are less accurate reflections of the places he had seen than illuminating distortions and metaphors for central themes. Abroad, then, in this essay is less actual place than a trope created by Conrad's imagination and a process by which the imagination travels to a world elsewhere. In Conrad's work, it is both the fictional other that Conrad seeks and the creative place that both confirms and questions his feeling of exile and marginality.

II

Conrad's life in the British Merchant Marine and his travels to the East played an important role in his fiction. Since he had actually sailed on a ship named the *Narcissus* in 1884, he could draw upon romantic memories of a successfully completed voyage at a time when his creative impulses were stifled by doubts.[3] Conrad also fused stories he had heard—such as the source story for the tale of James Wait—with his own journeys. Prior to writing *The Nigger of the "Narcissus"* (1897), Conrad sought an appropriate plot structure and point of view with which to organize his subject matter. Imagining the voyage of the *Narcissus* as a structural principle, he overcame writing paralysis. After he had committed himself completely to literature ("Only literature remains to me as a means of existence"), he was bogged down with the early version of "The Rescuer" (March 10, 1896; *Life and Letters* 1.185). Agonizing about his inability to make progress on "The Rescuer," he wrote to Garnett: "Now I've got all my people together I don't know what to do with them. The progressive episodes of the story *will* not emerge from the chaos of my sensations. I feel nothing clearly" (June 19, 1896; *Life and Letters* 1.192).[4] The imagined voyage of the *Narcissus* became at once the material for a plot to examine ethical and political questions of fundamental importance to Conrad and a private metaphor for the process of creating significance. Frustrated with his inability to write, the voyage of the *Narcissus* provided Conrad with an imaginative escape to the space and time of past successes.

"Youth" (1898), too, is a story that transforms the ingredients of a former sea voyage—in this case, his position as a second mate on the *Palestine* in 1881–82. Based on his command of the *Otago* in 1888—Conrad's only command—*The Shadow-Line* (1916) explores the difference between merely practicing skills and providing leadership to a community.[5] In contrast, the seemingly similar "The Secret Sharer" (1910), which is also based in part on his first command and in part on what Conrad had learned of an 1880 incident aboard the *Cutty Sark*, emphasized the captain-narrator's personal psychological development, rather than his ability to occupy a position in terms of standards established by maritime tradition. By fulfilling the moral requirements of a clearly defined position, the captain-narrator fulfills himself; he overcomes ennui, anxiety, and anomie and merges his psychological life with the demands of the external world. Thus the sea voyage— with its clearly defined beginning and ending, its movement through time toward a destination, its separation from other experiences, and explicit requirements that must be fulfilled by the crewmen and officers—provided a correlative *within Conrad's own experience and imagination* for the kind of significant plot that he sought.

III

Conrad visited the Malay Archipelago while sailing as first mate on the steamer *Vidar* (1887–88). Conrad's first two novels, *Almayer's Folly* (1895) and *An Outcast of the Islands* (1896), reflect his 1895–96 state of mind and reveal his values.[6] In these early novels Conrad tests and refines themes and techniques that he will use in his subsequent fiction. In what will become characteristic of Conrad's early works, he uses material for his fiction from his own adventures. He not only draws upon his Malay experience, but bases the title character of his first novel on a man he actually knew. While these two novels seem to be about remote events, they actually dramatize his central concerns.

Sambir, the setting for *Almayer's Folly* and *An Outcast of the Islands*, is the first of Conrad's distorted and intensified settings. Like the Congo in *Heart of Darkness* (1902) and Patusan in *Lord Jim* (1900), Sambir becomes a metaphor for actions that occur there. It is also a projection of Conrad's state of mind as it appears in his 1894–96 letters: exhaustion and ennui alternate with spasmodic energy.[7] Conrad's narrator is in the process of creating a myth out of Sambir, but the process is never quite completed. Like Hardy's Egdon Heath, Sambir is an inchoate form that can be controlled neither by man's endeavors nor by his imagination. The demonic energy that seethes within the forests is a catalyst for the perverse sexuality of the white people and their subsequent moral deterioration. With its "mud soft and black, hiding fever, rottenness, and evil under its level and glazed surface," Sambir refutes the Romantic myth that beyond civilization lie idyllic cultures in a state of innocence (325–26). Sambir's river, the Pantai, is a prototype for the Congo; the atavistic influence it casts upon white men, drawing out long repressed and atrophied libidinous energies, anticipates the Congo's effect on Kurtz. Sambir's primordial jungle comments on the illusion shared by Dain and Nina, as well as by Willems and Aissa, that passionate love can transform the world. Sambir's tropical setting seems to be dominated by the processes of death and destruction, and the jungle's uncontrollable fecundity moves toward chaos rather than toward order or evolution. The dominance of the Pantai and the forest implies that Conrad's cosmos is as indifferent to man's aspirations as the cosmos of his contemporary Hardy, whose *Jude the Obscure* was published in 1895.

Had Conrad not gone on to write *Heart of Darkness*, we might be more attentive to the extent to which Sambir embodied Conrad's nightmare of various kinds of moral degeneracy and how it is for him a grim Dantesque vision of damnation. Upon the anarchical and primordial Sambir, man seeks to impose his order. Lacking wife and parents, and bereft in England of any family ties, Conrad proposes family and personal relationships as an alternative to the greed and hypocrisy that dominate Sambir life. Throughout Conrad's early work, he dramatizes the search for someone to legitimize one's activities by an empathetic response to one's motives and feelings. We see this in Jim's need to be understood by Marlow and Marlow's to be understood by his audiences, as well as Conrad's desperate early letters to Garnett.

In his exotic Malay landscapes, Conrad created mindscapes of his own concerns. As an orphan who felt guilty for betraying his personal and national paternal heritage by living in England, Conrad was concerned from the outset with the relationship between parent and child. In the Sambir novels each person seems to require someone else to share his confidence. This takes the form of a search for the missing family. Almayer and Willems lack a father and seek to compensate for the absence of someone in whom to confide. The Malays' search for the restored family parallels that of the white protagonists: Omar is a father figure to Babalatchi, and the latter plays that role for Lakamba.

Conrad's fascination with human decadence, begun in the Malay novels with Almayer and Willems and continued in "The Idiots" (1896), is the subject of his first Congo story, the powerful and underestimated "An Outpost of Progress" (1898), which was written during an interlude from *The Rescue* (1920). Because he was bogged down on *The Rescue*, writing a story in which the moral distinctions were clear and in which he was in complete control of his materials was extremely important to him. Conrad's letter to Fisher T. Unwin makes clear that the story is in part an intense response to his 1890 Congo experience: "All the bitterness of those days, all my puzzled wonder as to meaning of all I saw—all my indignation at masquerading *philanthropy* have been with me again while I wrote" (emphasis mine).[8]

In "An Outpost of Progress,"Conrad examines his 1890 Congo journey— the source of *Heart of Darkness*—for the first time. When we examine Conrad's 1890 Congo Diary—reprinted in *Last Essays*—we see how Conrad's anterior reality informs a text, and how the imagination creates the reality of place to meet the thematic needs of a text. Note the diary entries:

Friday, 4th of July. . . . Saw another dead body lying by a path in an attitude of meditative repose. At night when the moon rose heard shouts and drumming in distant villages. Passed a bad night.[9]

Monday, 7th July. . . . Hot, thirsty and tired. At eleven arrived on the mket place. About 200 people. No water. No camp place. After remaining for one hour left in search of a resting place. Row with carriers. No water. . . . Sun heavy. Wretched.[10]

The phases of two white men's—Kayerts's and Carlier's—degeneration reflect Conrad's profound disillusionment from his own experience in the Congo. Despite their lip service to idealism, by the second day they have already ceased trying to export their civil service version of European civilization. In fact, the outpost of progress quickly becomes an outpost of savagery. Rather than being agents of change, these men are changed: like Kurtz in *Heart of Darkness* they gradually regress to savagery. But while Kurtz actually *renounces* civilized values and boldly practices "unspeakable rites," Kayerts and Carlier forget their ideals and drift into anomie. If Eliot excluded Kurtz from his category of "hollow men" because Kurtz chose evil, he could well have had Kayerts and Carlier in mind as the hollow men who did not will their fate. In progressive stages, the trappings of civilization crumble. First, Kayerts and Carlier abandon their attempts to improve their outpost. Then, they abdicate the vestiges of their morality when they accede to Makola's trading of slaves for ivory. Finally, they revert to complete savagery when Kayerts, after he thinks that Carlier intends to do the same to him, murders his companion.

That Marlow, the principal narrator of "Youth," *Heart of Darkness, Lord Jim*, and, later, *Chance*, is a vessel for some of Conrad's doubts and anxieties and for defining the problems that made his own life difficult is clear not only from his 1890 Congo Diary and the 1890 correspondence with Madame Poradowska, but, even more so, from the letters of the 1897–99 period. *Heart of Darkness* expresses his attitude toward the 1890 voyage to the Congo Free State and his response to the imperialistic excesses of Leopold II of Belgium.

The subject of *Heart of Darkness* is primarily Marlow, but the presence of Conrad is deeply engraved on every scene. Conrad dramatized Marlow's efforts to narrate his experience at a time when he himself was anxious that he might not be able to fulfill his artistic credo—as presented in the Preface to *The Nigger of the "Narcissus"*—of making other men *see*. Conrad transfers to Marlow the agonizing self-doubt about his ability to transform personal impressions into a significant tale. Marlow's effort to come to terms with the Congo experience, especially Kurtz, is the crucial activity that engaged Conrad's imagination. Marlow's consciousness is the arena of the tale, and the interaction between his verbal behavior—his effort to find the appropriate words—and his memory is as much the *agon* as his Congo journey. Both the epistemological quest for a context or perspective with which to interpret the experience and the semiological quest to discover the signs and symbols that make the experience intelligible are central to the tale.

The Congo experience had plunged Marlow into doubt and confusion. Marlow's experience in the Congo invalidated his naive belief that civilization equalled progress. As Marlow engages in an introspective monologue, the catalyst for which is his recognition that the Thames, too, contained the same potential darkness for the Romans as the Congo does for him; he recalls how he had discovered the pretensions of European civilizations. The equation of the Roman voyage up the Thames and Marlow's up the Congo suggests an important parallel. That Marlow says "this also has been"—not "this also was"—"one of the dark places on the earth" denies the idea of humanity's progressive evolution, still a widely held view in the 1890s, by showing that the manifestation of barbaric impulses is a continuous possibility. The essential nature of Europeans and natives is the same: "The mind is capable of anything—because everything is in it, all the past as well as all the future" (96). Conrad stresses that illusions are not only a defense against reversion to primitive life, but the basis of civilization.

One of Conrad's characteristic themes is the relationship between experience and memory. When memory imaginatively transforms our actual experience—the experience of the self that has been "abroad"—what do we leave behind? Although every event is informed by his present attitudes, Marlow's meditation follows the order of the original experience until he reaches the circumstances surrounding his first meeting with Kurtz. He desperately wants to believe that his journey into the atavistic Congo and his climactic encounter with Kurtz have broken down his personality only to prepare him for a new, broader integration and a deeper understanding of his relationship with—and responsibility to—other people. Marlow defers recounting the meeting with Kurtz in order to leap ahead to his meeting with the Intended, to comment on Kurtz's megalomania, and to relate how he saw the shrunken heads. He had difficulty recollecting his impression of the more gruesome details of his experience. Except for the shrunken heads, he contents himself with merely alluding to "subtle horrors" and "unspeakable rites." He claims he had almost expected to discover symptoms of atavistic behavior in his journey "back to the earliest beginnings of the world." (He specifically says, "I was not so shocked as you may think" [130].) Yet, now that Marlow is back within the civilized world, he recoils from the grotesque memories.

Marlow's journey from Europe to the Congo helped prepare him to sympathize with Kurtz. From the outset he was offended by the standards and perspectives of the European Imperialists, and gradually he began to sympathize with the natives against the predatory colonialists. As an idle passenger on a boat taking him to the Congo, he caught glimpses of the inanity he later encountered as an involved participant. Even then, he saw the fatuity of the "civilized" French man-of-war's shelling the bush. Marlow invests Kurtz with values that fulfill his own need to embody his threat of the jungle into one tangible creature. While Kurtz, the man who seemed to embody all the accomplishments of civilization, has reverted to savagery, the cannibals have some semblance of the "restraint" that makes civilization possible. While scholars have argued possible sources for Kurtz—and one can make a case for the parallel to the journalist/explorer Henry Morton Stanley—it is best to see Kurtz as a composite figure who symbolizes the pretensions of Imperialism. Kurtz is a poet, painter, musician, journalist, potential political leader, a "universal genius" of Europe, and yet once he traveled to a place where the earliest beginnings of the world still survived, the wilderness awakened "brutal instincts" and "monstrous passions." If Kurtz is considered the center of the "heart of darkness," the business of following Kurtz and winning the "struggle" enables Marlow to believe that he had conquered a symbol of the atavistic, debilitating effects of the jungle.

IV

Conrad's novels about politics have been viewed both as nihilistic statements

and dramatizations of a political vision. While the subject of these novels—
Nostromo (1904), *The Secret Agent* (1907), and *Under Western Eyes*
(1911)—is often politics, their values are not political. The novels affirm the
primacy of family, the sanctity of the individual, the value of love, and the
importance of sympathy and understanding in human relations. Conrad's
friendships with Cunninghame Graham, Wells, and Shaw show that he put
personal relationships before political ideology. His concern for the working
class derives not from political theory but from his experience as a seaman
and from his imaginative response to the miseries of others. Conrad's
humanism informs his political vision. In his political writings, it is the
abstractions upholding private virtues that carry conviction. He wrote in
"Autocracy and War" (1905) that it was to "our sympathetic imagination"
that we must "look for the ultimate triumph of concord and justice"
(*Collections* 84). In "Autocracy and War" the paramount values threatened
by Russian autocracy are "dignity," "truth," "rectitude," and "all that is
faithful in human nature" (*Collections* 99). Thus, Russia is "a yawning
chasm open between East and West; a bottomless abyss that has swallowed
up every hope of mercy, every aspiration towards personal dignity, towards
freedom, towards knowledge, every ennobling desire of the heart, every
redeeming whisper of conscience" (*Collections* 100).

Nostromo shows how Conrad became a figure of capable imagination who
could use his reading and a brief visit to the West Indies in 1878 to create an
imaginary nation with its own history and landscape. The pain of exile and
the feeling of marginality are important in Conrad. It is not necessary to
agree with Jocelyn Baines that Costaguana is a disguised version of Poland
to understand *Nostromo* as a sublimated act of self-justification on Conrad's
part.[11] For Conrad was deeply troubled over accusations that he had
abandoned Poland. He suspected his motives for settling in England and
turning his back upon both his country and his family heritage. Once he left
the sea and became a writer, the justification that Poland lacked the facilities
for his chosen career was difficult to sustain.

It may be that *Nostromo* also reflects Conrad's subconscious resentment
toward a father who neglected family for politics and ultimately left him an
orphan, after inflicting exile, disgrace, and economic hardship upon his
family. If *Nostromo* is subconsciously written to atone for Conrad's turning
his back on his father's tradition, it is hardly surprising that the catalytic act
that generates the novel's plot and the decisive act in the history of Costaguana is Gould's return to the land of his father's defeat for the purpose of
reviving the mine. By castigating the possibility of change through politics,
Conrad convinced himself of the rectitude of his own decision to desert
Poland, to which his father had made a complete political commitment.
Given his father's zealotry and Conrad's consciousness of it, it is hardly too

181

much to say that politics becomes a paternal abstraction to which Conrad must atone, palliate, and explain himself. The means of palliation are his political novels. *Nostromo* justifies the choice of personal fulfillment over political involvement because it shows politics as a maelstrom that destroys those it touches and, more importantly, shows that one inevitably surrenders a crucial part of one's personality when one commits oneself to ideology.

Under Western Eyes depends on a juxtaposition of Geneva and Russia. Conrad had a deep abiding hatred for Russia because he believed it had exploited and terrorized Poland and was responsible for the death of his parents. Conrad followed events in Russia and knew, through his friends the Garnett family, Russian refugees and revolutionaries. In the author's note he wrote that the plot and characters "owe their existence to no special experience but to the general knowledge of the condition of Russia."

In Conrad's version of Russia, autocratic politics create a world in which personal lives are distorted by the political abstractions served by proponents and antagonists. Each of the Russians creates for himself the fiction of a receptive counterpart who understands his every thought and feeling. Russia finally emerges as primitive and atavistic, a kind of European version of the Congo, where possibilities exist that have all but been discarded by Western countries.

Conrad's Geneva is a civilization where the libidinous energies and the atavistic impulses may be squelched, but violence and anarchy are under control. Conrad had visited Geneva in 1907 when he had taken his son Borys there for hydropathic treatment of suspected tuberculosis, the disease that both Conrad's parents had died of. He had been there once before in the 1890s on his own behalf. In a note in crude copy, he had written that he had been "induced to write this novel by something told me by a man whom I met in Geneva many years ago (Razumov's fate)."[12] It is very much to the point that the people, other than the revolutionaries, who reside in Geneva are engaged in shopkeeping, teaching, picnicking, walking; and that these quite ordinary activities can take place in Geneva, unlike in Russia, without bombs and intimidation. Geneva may have its materialistic aspect, epitomized by the rather tasteless Chateau Borel that now stands abandoned by its absentee owners, but it makes possible the cultivation of personal affections and the fulfillment of private aspirations, which the autocratic and violent Russian world blunts.

The narrator continually tests and redefines qualities that he associates with Russia and Geneva until, finally, he establishes the *moral* superiority of Western life. Like the narrator's fascination with Russian behavior, his repressed romantic interest in Natalie, and his imaginative excitement as he describes Razumov's self-flagellation in physical images, his muted dissatisfaction with Geneva indicates a repressed and sublimated longing for more intense experience than his lonely bachelor existence provides. Moreover, Geneva's willingness to accommodate the callousness and irrationality of the refugee revolutionary community within its midst offends his sense of morality. If, Conrad implies, the self-discipline of Western life has its cost in passion, it is nevertheless true that benign government gives people the choice of whether to write fictions, teach languages, or even pursue political visions. Geneva is a drab and pedestrian depiction of political stability, but it still remains a place where such a figure as the narrator may combine a highly civilized conscience with an individuality that, in its insistence on self-denigration, approaches the idiosyncratic and quirky.

In contrast to the narrator, who intuitively transforms every incident in his life into a matter of conscience, the Russians see their private lives in terms of a vague historical perspective. Thus the narrator's excerpts from Razumov's diary, Peter's autobiographical volume, and Tekla's life expand the novel's spatial-temporal dimensions. But the movement of the novel alternates between the personal, limited, and subjective perspective of the narrator and the vast, impersonal immensity of Russia with its countless anonymous citizens suffering misery that can barely be implied:

> Razumov received an almost physical impression of endless space and of countless millions.
>
> He responded to it with the readiness of a Russian who is born to an inheritance of space and numbers. Under the sumptuous immensity of the sky, the snow covered the endless forests, the frozen rivers, the plains of an immense country, obliterating the landmarks, the accidents of the ground, levelling everything under its uniform whiteness, like a monstrous blank page awaiting the record of an inconceivable history. (33)

Conrad deliberately depicts Geneva as tediously geometric and rather claustrophobic. Razumov is contemptuous of its decorum; he regards the view of the lake as "the very perfection of mediocrity attained at last after centuries of toil and culture" (203). While Russian political zealots such as Peter and Sophia—and Mikulin and General T—speak of national destiny and political ideals, the narrator's life is concerned with personal relationships in the "free, independent, and democratic" city of Geneva. The narrator speaks condescendingly of the "precise" and "orderly" Genevan landscape, but the very precision of the narrator's description, as well as his personal subjective response to place, implicitly criticizes the unlimited, amoral space of Russia. The novel confirms the value of the mind's own interior space, personal communication, and private relationships; it rejects historical and geographical explanations that seek to place moral responsibility beyond the individual conscience. The humanity and perspicacity that the narrator brings to his reminiscence "contain" and undermine the Russian conception of vast objective space that resists man's effort to domesticate it.

V

Conrad's neglected masterpiece, *The Rover* (1923), deals more with his reshaping history to fit a dream of a heroic, self-sacrificing death and less with the actual events of the Napoleanic war. In 1924, some months before he died, Conrad spoke of *The Rover* in terms that suggested its special importance to him: "I have wanted for a long time to do a seaman's 'return' (before my own departure)" (February 22, 1924; *Life and Letters* 2.339). Peyrol's desire, in his final voyage, to merge his destiny with that of his nation may reflect Conrad's desire, as he approached death, to contribute meaningfully to Poland's destiny. His fantasy of a significant political act is embodied in Peyrol. If, like Nabokov's, Conrad's life was embodied in his imagination, he was never comfortable that he had turned his back on politics and the heritage of his father, whom he recalled as an idealistic patriot. The novel's title also refers to himself, the twice-transplanted alien who finally found a home in England and no longer felt himself something of an outsider. Peyrol re-creates himself at fifty-eight when circumstances connive with his own weariness to deprive him of his past; he creates a new identity just as surely as a younger Conrad did when he left Poland to go to sea and an older Conrad did later, when he turned from the sea to a writing career.

The Rover combines Conrad's fantasy of retreat with his lifelong fantasy of a heroic return home. (Neither his first visit to Poland in 1890, nor his second at the outbreak of the First World War, quite fulfilled his fantasy.) *The Rover* associates Peyrol's return with Conrad's own romantic desire to return to his past. In *A Personal Record*, writing of his first return to Poland, Conrad wrote how the faces "were as familiar to me as though I had known them from childhood, and my childhood were a matter of the day before yesterday" (27). Similarly, upon arriving in revolutionary France, from which he had long absented himself while pursuing a career as an adventurer, Peyrol is struck by the parallel between himself and the people he encounters, including even the cripple. Gradually he feels that he belongs to France, represented in his mind by the tiny coastal hamlet in which he lives and the people he knows there:

> The disinherited soul of that rover ranging for so many years a lawless ocean . . . had come back to its crag, circling like a great sea bird in the

dusk and longing for a great sea victory for its people: that inland multitude of which Peyrol knew nothing except the few individuals on that peninsula cut off from the rest of the land by the dead water of a salt lagoon. (142)

Peyrol embodies Conrad's lifelong relationship with Poland—the country he had left—and his desire to return to the land of the parents who had died while he was a young child. Finally, Peyrol underlines the theme of this essay: that Conrad's exile and travels were sources not only of physical settings, but imaginary places that were inextricably related to his own psyche. For Conrad—that figure of capable imagination—abroad was reshaped by his imaginatively transformed places, history, and politics for the purpose of emphasizing his themes and values.

NOTES

Quotations from Conrad's writings are from the Kent Edition of his works (Garden City, NY: 1926); his letters are quoted from *Joseph Conrad: Life and Letters*, 2 vols, ed. G. Jean-Aubry (Garden City, NY: 1927), unless otherwise noted.

1. *Poems by Wallace Stevens*, selected by Samuel French Morse (New York: 1959), 102.
2. M.M. Bakhtin, *The Dialogic Imagination*, ed. Michael Holquist, trans. Caryl Emerson and Michael Holquist (Austin, TX: 1981), 84.
3. See Jocelyn Baines, *Joseph Conrad: A Critical Biography* (New York: 1960), 75–77.
4. "The Rescuer" later became *The Rescue*. The letters are from G. Jean-Aubry's *Joseph Conrad: Life and Letters*, (abbreviated *Life and Letters*). The best biographical source is Zazislaw Nadjer's *Joseph Conrad: A Chronicle* (New Brunswick, NJ: 1983).
5. Conrad had written in Richard Curle's copy of the novel,

   This story had been in my mind for some years. Originally I used to think of it under the name of *First Command*. When I managed in the second year of war to concentrate my mind sufficiently to begin working I turned to this subject as the easiest. But in consequence of my changed mental attitude to it, it became *The Shadow-Line*. (Richard Curle, *The Last Twelve Years of Joseph Conrad* [London: 1928]; quoted in Karl, *Conrad: The Three Lives* [New York: 1979], 770.)

6. I use *Folly* to indicate *Almayer's Folly*, *Outcast* to indicate *An Outcast of the Islands*, and *Collections* to indicate Conrad's *Collections of Essays and Notes on Life and Letters*. Although written first, *Almayer's Folly* takes place twenty years

after *An Outcast of the Islands*. Conrad completed *Almayer's Folly* in 1894 and *An Outcast of the Islands* in 1895; the dates in parentheses are publication dates.

7. See G. Jean Aubry, *Joseph Conrad: Life and Letters*. Also John A. Gee and Paul J. Sturm, *Letters of Joseph Conrad to Marguerite Poradowska*, 1890–1920 (New Haven, CT: 1940); see, for example, 63, 82, 86, 88.

8. 1896 letter to Fisher T. Unwin, quoted by John Dozier Gordan, *Joseph Conrad: The Making of a Novelist* (Cambridge: 1940), 242.

9. Quoted in Jocelyn Baines, *Joseph Conrad: A Critical Biography* (New York: 1967), 115.

10. Quoted in Baines, 116.

11. See Baines, 313–14.

12. Quoted in Baines, 370.

"IRRECONCILABLE DIFFERENCES": ENGLAND AS AN "UNDISCOVERED COUNTRY" IN CONRAD'S "AMY FOSTER"

Keith Carrabine, University of Kent

On April 2, 1902, from "Pent Farm," on Romney Marsh in Kent, where he had lodged since September 1898, Joseph Conrad described "Amy Foster" to his French translator:[1]

> Story of an Austro-Polish highlander emigrating to America who is shipwrecked on the English coast. He roams around the countryside. No one understands him. He is tracked down like a wild beast. A farmer takes him in. He marries a good fool of a girl, a villager. But he persists in speaking his language to her and their child. That annoys his wife. She begins to fear him. He falls ill. She watches over him. He is thirsty and asks for a drink—but in his own language rather than hers. She doesn't understand. He can no longer understand why she doesn't stir. He becomes angry, he gets up—his alien noises frighten her and she runs away, carrying the child. He dies without ever understanding. . . . Idea: the essential difference of the races.

This "Idea," with its attendant offspring—the differences between the various races' histories, cultures, values, metaphysics, and not least, languages—had been Conrad's great subject from the opening line of his first novel *Almayer's Folly* (1895), when Mrs Almayer, a Malay, yelled "Kaspar! Makan!"—and startled Almayer, a Dutch trader, from his dream of a "splendid future" in Amsterdam.[2] The Malayan setting of *Almayer*, as well as of *An Outcast of the Islands* (1896), "The Lagoon" and "Karain" (*Tales of Unrest* [1898]), and *Lord Jim* (1900), as many commentators have noticed, solved one great problem for the Polish "exile and wanderer" born Józef Teodor Konrad Korzeniowski (coat of arms Nalecz) in Berdyichev, Podolia, in the Russian Ukraine—namely "his lack of a common cultural background with his readers." Hence the exotic Eastern seas and the Malay Archipelago provided in Najder's words "a safe ground, a concrete basis, to enable him to write with a certain amount of authority yet without the risk of discussing matters with which his English readers were too familiar."[3]

It still has not been sufficiently recognized that Conrad's "Idea," with all its ramifications, was the inevitable subject and expression of his vision of the artist's task and of man's fate—most memorably formulated in his great letter of August 2, 1901, written barely a month after the completion of "Amy Foster":

> The only legitimate basis of creative work lies in the courageous recognition of all the irreconcilable antagonisms that make our life so enigmatic, so burdensome, so fascinating, so dangerous—so full of hope. They exist! And this is the only fundamental truth of fiction.[4]

Because Conrad believed with Flaubert that "the *whole* of the truth lies in the presentation," his life-long task as an artist was the search for forms correlative to his austere vision of "irreconcilable antagonisms."[5]

In the case of "Amy Foster," Conrad reworks fictional strategies first employed in "Youth" and *Heart of Darkness*. Thus the tale is told in England by an unnamed frame-narrator, who introduces and comments upon the main story, which is narrated by a Marlovian figure called Dr. Kennedy. *Heart of Darkness*, however, like his earlier fiction, depicts "the essential differences of the races" in a far-flung region of the world, but "Amy Foster," in a unique experiment, places a Pole at the center of a fiction set in Romney Marsh, Kent, England. As an examination of Conrad's depiction of landscape, of his manipulation of the interpenetrating perspectives and voices of his two English narrators, of his Polish protagonist, and of the Kentish villagers will show, this bold transfer of location and startling switch of the races enabled him to rehearse the nodal idea of his fiction and, courageously, to present England itself as "an undiscovered country."

I

The frame-narrator in the first fifth of "Amy Foster" performs three major functions: he describes the setting of the story, he introduces us to his friend, Dr. Kennedy, and he registers, as the first listener, the kinds of responses Conrad seeks to elicit from his readers. The story begins with a description of Kennedy's home "in Colebrook":

> The high ground rising abruptly behind the red roofs of the little town crowds the quaint High Street against the wall which defends it from the sea. Beyond the sea-wall there curves for miles in a vast and regular sweep the barren beach of shingle, with the village of Brenzett standing out darkly across the water, a spire in a clump of trees; and still further out the perpendicular column of a lighthouse, looking in the distance no bigger than a lead-pencil, marks the vanishing-point of the land. The country at the back of Brenzett is low and flat; but the bay is fairly well sheltered from the seas, and occasionally a big ship, windbound or through stress of weather, makes use of the anchoring ground a mile and a half due north from you as you stand at the back door of the "Ship Inn" in Brenzett. A dilapidated windmill near by, lifting its shattered arms from a mound no loftier than a rubbish-heap, and a Martello tower squatting at the water's edge half a mile to the south of the Coastguard cottages, are familiar to the skippers of small craft. (105)

The High Street may be "quaint" but it is *defended* "from the sea" by "the wall." Yanko will of course emerge mysteriously "from there" (111). Hence this initial description foreshadows how the threat of his otherness will be resisted by the walls of custom and prejudice, which, the subtle switch of perspective suggests, are as essential to the townsfolk's ability to survive as are the walls and houses they have built to shield themselves from the barrenness of their coast and from the destructive force of the sea. Again "a spire in a clump of trees," a "dilapidated windmill" with its "shattered arms," and "a mound no loftier than a rubbish-heap," like the station viewed through Marlow's glasses ("A long decaying building on the summit was half-buried in the high grass; the large holes gaped black from afar" [*Heart of Darkness* 121]) are suggestive of the potential decay inherent in all man-made designs and at the heart of civilization itself. Thus this seemingly straightforward description actually prepares the reader to accept, with more ready sympathy and understanding, Yanko's sense of Kent as a mysterious and hostile space.

The interrelation between the setting, the narration, and the events of the tale is further illustrated in the frame-narrator's powerful description of the sunset, which follows immediately upon Kennedy's musings about "the surprises of the imagination" that inspired Amy's initial love for Yanko and about the "tragedies . . . arising from the irreconcilable differences and from that fear of the Incomprehensible that hangs over all our heads":

> the rim of the sun, all red in a speckless sky, touched familiarly the smooth top of a ploughed rise near the road as I had seen it times innumerable touch the distant horizon of the sea. The uniform brownness of the harrowed field glowed with a rose tinge, as though the powdered clods had sweated out in minute pearls of blood the toil of uncounted ploughmen." (108)

Under the spell of Kennedy's melancholy, universal musings, the familiar and the "uniform" (in a brilliant reversal) are destabilized, and the painterly "powdered clods" are rendered as being as much insatiable for, and as threatening to, human aspirations as either "the intense work of tropical nature" in *Almayer's Folly* (71) or "the living trees, lashed together" of the Congo jungle in *Heart of Darkness* (101). And as the Darwinian "plants" in the former "shooting upward, entwined, interlaced in inextricable confusion," "violently subvert" in Ian Watt's formulation "the conventional assumptions of popular romance" so "the minute pearls of blood" (more poignantly as befits the tale) subvert the assumptions of the pastoral mode.[6] Similarly Kennedy's mournful assessment of Amy's awakening from the "possession" of "love" into "a fear resembling the unaccountable terror of a brute" imbues the frame-narrator's sense that "a penetrating sadness . . . disengaged itself from the silence of the fields. The men we met [returning from work] walked past, slow, unsmiling, with downcast eyes, as if the melancholy of an over-burdened earth had weighted their feet, bowed their shoulders, borne down their glances" (110). The "earth" is simultaneously overburdened by their "toil" and by their heavy natures: yet, conversely, as Kennedy observes, their unceasing obligation to harrow the land is "a curse" (with all its biblical resonances) that they cannot avoid—a "curse" that loads "their very hearts . . . with chains" (111).

At this point we are one fifth of the way into "Amy Foster" and Yanko finally appears. Because we are immediately warned by Kennedy that "He was so different from the mankind around," we fear that this "castaway . . . washed ashore here in a storm" will find the uncouth and enchained villagers a greater threat to his security than the destructive element itself. Enchained and uncouth, the villagers are "taken . . . out" of their "knowledge" by Yanko's mysterious emergence from "out there" as surely as he is out of his "knowledge," when he is cast away by the sea upon "the barren beach of shingle."

As in *Heart of Darkness* the unnamed narrator informs us that Marlow was "a wanderer" fond of describing "inconclusive experiences" (48), so his counterpart in "Amy Foster" introduces Dr. Kennedy and releases details that prepare us for the kind of tale he will tell:

He had begun life as surgeon in the Navy, and afterwards had been the companion of a famous traveller, in the days when there were continents with unexplored interiors. His papers on the fauna and flora made him known to scientific societies. And now he had come to a country practice—from choice. The penetrating power of his mind, acting like a corrosive fluid, had destroyed his ambition, I fancy. His intelligence is of a scientific order, of an investigating habit, and of that unappeasable curiosity which believes that there is a particle of general truth in every mystery. (106)

Galsworthy, commenting on Conrad's appreciation of "the super-subtle, the ultra-civilised in James's fiction," went on to note: "And yet there is not . . . a single portrait in his gallery of a really subtle English type, for Marlowe though English in name is not so in nature."[7] Neither we might add is Kennedy, whose "nature," like Marlow's, has been formed abroad. It is precisely because Kennedy on the one hand has experienced "continents with unexplored interiors" and has therefore "felt profoundly" his "own strangeness,"[8] and because he has, on the other, an "intelligence of a scientific order," that he is able to empathize with both Yanko's and the villagers' predicaments, and to classify "the particle of a general truth" in this terrible tale of "irreconcilable differences." (And, as the last part of this essay will show, Kennedy and Marlow are also un-English because their natures and their visions partake of their creator's Polishness.)

Kennedy's un-English subtleness is most fully revealed in the pivotal meditation, which follows upon the last significant interruption of the frame-narrator ("Kennedy's voice . . . passed through the wide casement, to vanish outside in a chill and sumptuous stillness"):

The relations of shipwrecks in the olden time tell us of much suffering. Often the castaways were only saved from drowning to die miserably from starvation on a barren coast; others suffered violent death or else slavery, passing through years of precarious existence with people to whom their strangeness was an object of suspicion, dislike or fear. We read about these things, and they are very pitiful. It is indeed hard upon a man to find himself a lost stranger, helpless, incomprehensible, and of a mysterious origin, in some obscure corner of the earth. Yet amongst all the adventurers shipwrecked in all the wild parts of the world, there is not one, it seems to me, that ever had to suffer a fate so simply tragic as the man I am speaking of, the most innocent of adventurers cast out by the sea in the bight of this bay, almost within sight from this very window. (113)

Kennedy appeals to and then refocuses the stereotypes of popular romance; his "penetrating" sympathy for "a lost stranger" transforms "the bight of this bay, almost within sight from this very window" into "some obscure corner of the earth"; concomitantly he realizes that "people to whom their [the castaways'] strangeness was an object of suspicion, dislike, or fear" do not necessarily inhabit "the wild parts of the world," but "this valley down to Brenzett and Colebrook and up to Darnford" which encompasses his "country practice." Kennedy is un-English at such moments because he combines Marlow's "confounded democratic quality of vision which may be better than total blindness, but has been of no advantage to me, I can assure you" (*Lord Jim* 94), *and* a tragic recognition of the "subtle poignancy arising from irreconcilable differences."

The "democratic quality of vision" of Conrad's narrators constitutes the necessary counterbalance to his tragic vision of "irreconcilable antagonisms." With an intensity unmatched in English literature since Wordsworth, Conrad's narrators articulate his awareness that

> labourers, fishermen, old wives, the common people of the earth and the sea shore, the inarticulate who live beyond the pale, people of obscure minds, of imperfect speech, whose complexity the heritage of mankind seems buried deep out of sight like a seed in hard ground waiting for the day, for the moment, for the fertilising touch of a sunray or a thunder shower, to pierce the shell of mud, to sprout grow and bear its own flower, tender or bizarre or ominous of terror and sorrow.[9]

They await, that is, "the fertilising touch" of the listener/storyteller who can transmute their "imperfect speech" into the language of art.

Conrad recognized that the lives of the inarticulate (whether victims or victimizers) challenged the capacities and the boundaries both of his humanity and of his art because, to adapt Marlow on Jim, "they complicate matters by being simple" (94). Hence a fine and typical Conradian fiction, like "Amy Foster," involves the reader in the narrator's hovering search for a form appropriate to this vision of "irreconcilable antagonisms"—a form that will simultaneously render the highest kind of justice to all "of obscure minds, of imperfect speech," whether natives of regions such as Kent and the Malay archipelago, or strangers in their midst like Yanko and Jim.

This form finds one of its most persuasive expressions in the whole of Conrad's fiction in the single sweep of Kennedy's tale (113–42), which continues, virtually without interruption, until this "simply tragic" story is completed.

II

As we have seen, the story is so organized that the filtering of Yanko's account of events is delayed. In narrative terms the frame-narrator's response to Amy as "a dull creature" triggers Kennedy's story of her "irresistible and fatal impulse" toward Yanko. And, because Amy's "possession" by Yanko and her subsequent awakening from "that enchantment . . . by a fear resembling the unaccountable terror of a brute" are both recorded before the castaway's story begins, it is as if his fate is entirely enclosed within hers. (Hence from this perspective the aptness of the story's title.) In terms of the tale's mood, however, Yanko's "simply tragic" career has become part of, and is generated by, "A sense of penetrating sadness, like that inspired by a grave strain of music," which disengages "itself from the silence of the fields" (110). Yanko's tale therefore, at the moment of its remembrance, is no longer bracketed within Amy's; rather, as a result of the joint pressure of Kennedy's reflections and his auditor's imaginative sympathy, it is now representative of the still sad music of humanity.[10]

Kennedy's account of how Yanko was "washed ashore here in a storm" and of his origins in "the eastern range of the Carpathians" (121) is split into two parts: the first (111–12) reinforces a key perspective—"for him who knew nothing of the earth, England was an undiscovered country"—and confirms that Kennedy's transmission is based on "his broken English that resembled curiously the speech of a young child." The second part (113–18) loops back to his origins.

Yanko's story begins:

He did not know the name of his ship . . . he did not even know that ships had names—"like Christian people"; . . . They were driven below into the 'tween deck and battened down from the very start. . . . It was very large, very cold, damp and sombre, with places in the manner of wooden boxes where people had to sleep one above another, and it kept on rocking all ways at once all the time . . . and . . . a great noise of wind went on outside and heavy blows fell—boom! boom! . . . It seemed always to be night in that place. . . .

Before that he had been travelling a long, long time on the iron track. He looked out of the window, which had wonderfully clear glass in it. . . . He gave me to understand that he had on his passage beheld uncounted multitudes of people. . . . Once he was made to get out of the carriage, and slept through a night on a bench in a house of bricks. . . . There was a roof over him, which seemed made of glass, and was so high that the tallest mountain-pine he had ever seen would have had room to grow under it. Steam-machines rolled in one end and out the other. People swarmed more than you can see on a feast—round the miraculous Holy Image in the yard of the Carmelite Convent down in the plains where, before he left his home, he drove his mother in a wooden cart . . . someone had told him it was called Berlin. . . .

They thought they were being taken to America straight away, but suddenly the steam-machine bumped against the side of a thing like a great house on the water. The walls were smooth and black, and there uprose, growing from the roof as it were, bare trees in the shape of crosses, extremely high . . . he had never seen a ship before. This was the ship that was going to swim all the way to America. (114–16)

It is typical of the balance of Conrad's "democratic vision" that this story, which will go on to record the Kentish villagers' and then Amy's ignorant reactions to Yanko's "difference, his strangeness" (137), first charts the strangeness that overpowers Yanko from the moment he leaves his mountain home. Yanko the naif knows neither "the name of his ship" nor the city where "he slept through the night." His vocabulary ("the steam-machines," "swim") and the religious and cultural indices he invokes ("Holy Images," "pines," "a wooden cart") to gauge his new experiences attest that Yanko is obliged to *decode* the strange new world he enters that has such people in it.[11] Simultaneously, as we do with Gulliver in Brobdignag, we also have to adjust our perspective to Yanko's. Thus we have to realign *our* perceptions ("bare trees in the shape of crosses") and thereby collaborate in, as well as "conceive . . . of an existence overshadowed, oppressed, by the everyday material appearances, as if by the vision of a nightmare."[12]

"Stupidity [incomprehension] in the novel" as Bakhtin has taught us, "is always polemical" (403); and Yanko's "broken speech" instills "a strangely penetrating power into the sound of the most familiar English words" (117).[13] It works, on one level, as a wry commentary upon the "swarm" of city life and as a bitter commentary upon the unscrupulous emigration officials who trade upon the credulities of "obscure minds" with false promises of "America . . . where true gold could be picked up on the ground." This phase of Yanko's story, however, ends abruptly:

No doubt he must have been abominably seasick and abominably un-happy—this soft and passionate adventurer, taken out of his knowledge, and feeling bitterly as he lay in his emigrant bunk his utter loneliness; for his was a highly sensitive nature. The next thing we know of him for certain is that he had been hiding in Hammond's pig-pound by the side of the road to Norton, six miles, as the crow flies, from the sea. Of these experiences he was unwilling to speak; they seemed to have seared into his soul a sombre sort of wonder and indignation (118).

This astonishing switch from "the emigrant bunk" to the "everyday materials" of "Hammond's pig-pound" obliges us, simultaneously, to share Yanko's disorientation and to wonder how such a pious Christian ended up "six miles . . . from the sea" in such degrading surroundings; and we are obliged to cope with the pointed adjustment of the tale's satirical edge as we are transferred from one kind of incarceration ("battened down in the manner of wooden boxes" in an "emigrant bunk") to another.

The reader's realignment is matched by a shift in the narrative mediation forced upon Kennedy by "experiences" of which Yanko "was unwilling to speak" (118). Hence from this moment on Yanko's voice is displaced and becomes one among, and circumscribed by, the cacophany of "imperfect speech" of "the common people of the earth and the sea shore" orchestrated by Kennedy. Moreover, because we have become attuned to both the nature and form of Yanko's "incomprehension" and to the narrators' *comprehension*, we view the villagers' and Amy's encounters with and responses to Yanko through both "the wonder and indignation" of the soul-seared castaway and the "simply tragic" vision of Kennedy's mediation.

From the moment Yanko lands on shore, his "piercingly strange words" (118) and "his flood of passionate speech" (126), together with his "miry" appearance, fill the villagers with the dread of inexplicable strangeness. Hence their "commonplace words" are void of charity and expressive of their fear of Yanko's otherness. He is variously "that dirty tramp," "a horrid-looking man," "a hairy sort of gypsy fellow," and "this troublesome lunatic." Finally:

> as the creature approached him, jabbering in a most discomposing manner, Smith (unaware that he was being addressed as "gracious lord," and adjured in God's name to afford food and shelter) kept on speaking gently but firmly to it and . . . by a sudden charge . . . bundled him headlong in to the wood-lodge, and instantly shut the bolt. . . . He had done his duty to the community by shutting up a wandering and probably dangerous maniac. (120–21)

The villagers' fierce ascriptions—"tramp," "gypsy," "lunatic," and "madman"—encode generations of ignorance, racial prejudice, and xenophobia akin to those of barbarous natives "in some obscure corner of the earth." And, as ever, in Conrad's fiction their names (like those of "enemies," "savages," and "brutes" applied to the natives in *Heart of Darkness* which sanction the horrors of Imperialism) exorcise the strange *and* enable them to exercise dominion over their world.[14] Because their language is *tribal*, and because they lack imagination, they fail to conceive that Yanko's "language that to our ears sounded so disturbing, so passionate and bizarre" (137) carries a human imprint: thus, again like the whites in Africa faced with the speech of the natives, they call Yanko's "flood of passionate speech," "jabbering," "babbling," and "senseless." Conrad's use of free indirect speech ("kept on speaking gently but firmly to it") carries a triple freight: it mocks the villagers' vigilant overreactions and their complacency, it renders Yanko's destiny truly pitiable, and it captures the commonplace bewilderment and genuine fear that led these people of "obscure minds" to treat a gentle Christian, craving the most basic human needs, as an animal. No wonder Yanko is "unwilling to speak" of his first contact with these inexplicably hostile English voices; and as Marlow, isolated on the French steamer "amongst all these men with whom I had no point of contact" greets "the voice of the surf" as "a positive pleasure, like the speech of a brother" (*Heart of Darkness* 61), so Yanko "welcomed" the "bleating in the darkness" of the sheep as "the first familiar sound he heard on these shores." All the human *voices* he encounters, from the "Jews" who claim to "talk to . . . the American Kaiser" (116) through the whole gamut of the villagers' outraged reactions, circumscribe his destiny. Finally Amy, whose "preliminary stammer" had seemed a badge of her sincerity, eventually surrenders to the ignorant, xenophobic judgments of her kin ("these foreigners behave very queerly to women sometimes" [135]), and (as Conrad told his French translator) terrified by Yanko's "alien noises," she abandons him.

The reactions of these "obscure minds" to Kennedy, who has "knowledge of unexplored interiors," are expressive of a universal *aversion* (in the root meaning of to "turn away"), generated by their inescapable "suspicion, dislike, or fear" of Yanko's "strangeness." Moreover, terribly, their reactions are stimulated by his increasingly desperate attempts to speak to them—in hope of making them *turn toward him*. To Yanko, however "taken out of his knowledge" and "innocent of heart and full of good will" ("gracious lord . . . in God's name") their behavior is as "outlandish," as inhuman, and as inexplicable to him as his is to them—reducing him, however, "very near insanity," and filling him with "rage, cold, hunger, amazement and despair" (121). Thus, through the prism of Yanko's startled perceptions, Conrad is enabled to rework systematically the reversal of roles and the wresting of the reader's imaginative sympathies achieved when Marlow views England, from the perspective of "a decent young [Roman] citizen," as "a swamp," where, "in some inland post" he feels "the utter savagery had closed round him" (*Heart of Darkness* 50). And, as in *The Secret Agent* (1907) which is set in London, the narration of "Amy Foster" is calculated to shock his English reading public into a recognition of the limitations of their social codes, of their insularity, and of the thinness of the protections they build against the invasion of the strange and alien.

The battle of perspectives and voices in this sequence typifies the remainder of the narrative. Kennedy views Yanko as pious and "innocent of heart and full of good will" (132), as "a woodland creature" and insists on "his very real beauty" (133). Kayerts and Carlier in "An Outpost of Progress" (*Tales of Unrest* 1898) are undone because they "are two perfectly insignificant and incapable individuals," stupid, lazy, and blind, and because, consequently, they cannot cope with their "contact with pure unmitigated savagery, with primitive nature and primitive man" (89). In marked contrast Yanko is undone, *despite* his human virtues and his grace, both by "the hostility of his *human* surroundings" (133, my emphasis), and, tragically, by his own (unexceptional) persistence (as Conrad noted) "in speaking his language to her and their child." Hence to Kennedy he becomes "an animal under a net," "a wild bird caught in a snare" (126).

To the community, however, his dress, his "swarthy complexion," "his rapid skimming walk" become "so many causes for offence" (132). For them "his foreignness had a peculiar and indelible stamp" (131–32). Yanko is "an excitable devil"; they dislike his "acrobat tricks in the tap room" and they give him "a black eye." The villagers regard his desire to marry Amy as "odious" and her father "preached it to his daughter that the fellow might ill-use her in some way" (135–36).

Yanko's naiveté is used to explore "what made them [the villagers] so hardhearted and their children so bold" (128). Through his simple pious vision England becomes, as it had for the Romans in *Heart of Darkness*, "one of the dark places of the earth." Hence we are made to experience the narrowness and spiritual impoverishment of the Protestant spirit. "He had approached them as a beggar it is true, he said; but in his country, even if they gave nothing, they spoke gently to beggars" (124). Again, unconscious of the biting irony of his words, Yanko observes, "If it hadn't been for the steel cross at Miss Swaffer's belt he would not, he confessed, have known whether he was in a Christian country at all" (129). Though "The rectory took much notice of him" the interest of the young ladies is dogmatically Protestant; and, rather than show concern for his human needs, they attempt "to prepare the grounds for his conversion" (131).

Yanko's own response of "overwhelming loneliness" to that "leaden sky of winter without sunshine" which, to his dejected vision, seems to have bred "faces of people from the other world—dead people" (129) echoes the frame-narrator's commentary upon the still sad music of humanity. The unequal collision of Yanko and the "dead people," like that between man and the land, and like that between Kurtz and the wilderness, ensures that Yanko "lying face down and his body in a puddle" (140) returns, again like Kurtz, who is buried "in a muddy hole," to the very "clods" from which all life once emerged.

Hence, whereas Kayerts unconsciously passes judgment on his own baffled, sordid life and destiny when he reflects on the death of Carlier ("that man died every day in thousands . . . and that . . . one death could not possibly make any difference . . . at least to a thinking creature" [115]), with Yanko's death we feel that a spontaneity and a vitality, absent from and unrecognized by the sluggish villagers, passes away from the earth.

III

Given that "Amy Foster" is the only tale Conrad set in England with a Pole as the central character, and given that it is one of his most powerful studies of alienation, it is not surprising that many Conradians regard it as "a work of spiritual autobiography embodying his ever-present feelings of loneliness, foreignness, and isolation, and his sense of exile from Poland."[15] Although it does not have to be, much of this biographical and psychoanalytical commentary is both reductive and overconfident about the conclusions we can draw when we note either the parallel elements in the careers of Yanko and his creator, or when we attempt to detect the Polish "figure" behind "the veil" of such works as "Amy Foster."[16] Moreover it has obscured the links between the historical, political, and cultural aspects of what Conrad called his "peculiar experience of race and family" and his nodal idea of "irreconcilable differences." (Author's Note to *Under Western Eyes* viii).

At the risk of oversimplification, the three most significant aspects of Conrad's Polish heritage are that he was born a subject of the vast Russian Empire—Poland's historic foe; that the region of his birth, the Ukraine, was multiethnic, multireligious, and multilingual; and that his own "immense parentage" suffered, like Razumov's (*Under Western Eyes*) from the throes of "internal dissensions" (11).

Before the partition of Poland in the late eighteenth century, the Ukraine had been part of the loose federation of the Old Polish Commonwealth. Two generations later, when Conrad was born, the population of the Ukraine consisted of a Russian (Orthodox Christian) bureaucracy and standing army, Polish (Roman Catholic) landowners and landlords, German (Protestant) colonists, a vast underclass of Ukrainian (Uniate Christian) serfs, and in the town of Berdichev, a large Jewish population. Crucially, as Pawel Hostoweic has noted, "each of these groups had their own truth, separated from others by religion, customs, tradition, profession," and, of course, by language."[17] The "essential difference of the races" in the Ukraine accounts in part for the bloody history of racial conflict in the eastern lands of the Old Polish Commonwealth from the late eighteenth century to the present. Moreover, the Polish loathing of "Muscovite barbarism" erupted in three insurrections (1794, 1830, 1863).

Born into a region characterized by schism and strife, it was Conrad's peculiar fate to endure, also, two opposed ideologies and ways of life within his own family. His father, Apollo Korzeniowski, was an ardent messianist and insurrectionist for "the great and godly cause of Poland," and was one of the leading spirits behind the abortive rebellion of 1863. Conrad and his mother accompanied Apollo into exile in Northern Russia in 1862. In marked contrast, the orphan Conrad's maternal uncle and guardian, Tadeusz Bobrowski, was a conciliator who scorned both the impracticality and the "inflammeable temperament" of his insurrectionist kin. He sought "the attainment of a fairly tolerable modus vivendi" with the loathed Russian autocracy "that would find its expression in a complete autonomy of the (Polish) kingdom, a recognition of our basic needs and, in time, an acknowledgement of certain national rights" to the territories of the Old Commonwealth.[18]

The complex relationship between Conrad's life and art is, I think, best approached through Bakhtin's richly suggestive account of Dostoevsky's career. Thus Conrad's multiethnic, multilingual, deeply divided Polish heritage, and his subsequent career as a mariner in the British Merchant Service (during which he was exposed to English, Dutch, and Belgian modes of Imperialism and to a great variety of cultures, truths, and languages), duplicates the cultural model Bakhtin constructs to account for Dostoevsky's resort to and development of the "polyphonic novel":

> the objective complexity, contradictoriness and multi-voicedness of Dostoevsky's epoch, the position of the déclassé intellectual and the social wanderer, his deep biographical and inner participation in the objective multi-leveledness of life and finally his gift for seeing the world in terms of interaction and coexistence—all this prepared the soil in which Dostoevsky's polyphonic novel was to grow.

Moreover, both Conrad and Dostoevsky experienced "the contradictory multi-levelness of their times" and understood "the extensive and well developed contradictions which coexisted among people" of different languages, faiths, and historical traditions, and which coexisted *within* these cultures and beliefs.[19]

It is surely not surprising, therefore, that as he matured as a writer, Conrad discovered that his art, like his heritage and life, was *grounded* in "irreconcilable differences." Indeed Conrad's unwavering commitment to the differing truths inherent in these "irreconcilable antagonisms" largely explains why his fiction and his narrators, such as Kennedy and Marlow, strike us as *foreign*. Moreover, Conrad's deeply divided heritage and his multilingual, multiethnic, multireligious culture begin to explain why his aesthetic of the novel anticipates Bakhtin's theories, and why his fictions

199

themselves correspond so closely to the latter's definitions of the genre's distinguishing characteristics. Thus "Amy Foster"'s multilayered, densely mediated, polyphonic form matches Bakhtin's definition of the novel "as a diversity of social speech types (sometimes even diversity of languages) and a diversity of individual voices, artistically organized." Hence, "Amy Foster" and Conrad's other major works prefigure and exemplify Bakhtin's insistence that "all languages of heteroglossia . . . are specific points of view on the world, forms for conceptualizing the world in words, specific world views, each characterized by its own objects, meanings and values. As such they all may be juxtaposed to one another, mutually supplement one another, contradict one another and be interrelated dialogically."[20]

They intertwine dialogically in, for instance, the narratives of Marlow and Kennedy. In this respect the "grave strain of music" in "Amy Foster" that "disengaged itself from the silence of the fields" expresses Conrad's strenuous aspiration to extract and to compose from the polyphony of "imperfect speech" a pure "strain" of "the magic suggestiveness of music—which is the art of arts" (Preface to The Nigger of the "Narcissus"). Yet, like Dostoevsky, Conrad was not sympathetic to the ultimate fusion or erasing of differences. Thus his narratives, like those of Dostoevsky, systematically utilize syncrisis (the juxtaposition of various points of view on a specific object [ivory, silver] or person [Yanko, Jim]) to show the irreducibility of the voices in his narratives—voices that are expressive of both the clamorous and rancorous humanity of the different races and communities and of the "irreconcilable differences" that "constitute the only fundamental truth of fiction."

Conrad's commitment to these "antagonisms" and his consequent sympathy for the huge range of voices canvassed and coexistent even in such a short narrative as "Amy Foster" help explain why he sounds so foreign placed next to such late-Victorian contemporaries as Wells, Galsworthy, and Bennett. The latter group lack the dialogic density and the multiplicity of voices inherent in the novel as a genre because, as Conrad told Wells, he failed to cast "a wide, a generous net . . . where indeed every sort of a fish would be welcome, appreciated and made use of," and because, as he told Galsworthy, "you hug your conceptions of right and wrong too closely." Like his contemporaries, Galsworthy wants "more skepticism . . . the tonic of minds—of life, the agent of truth—the way of art and salvation."[21]

For Conrad, as for Flaubert and James, "the only morality of art apart from subject" was "the truth that lies in the presentation," and to that end he strove to "leave enough to the imagination" of his readers.[22] Hence Conrad's fictions and particularly his endings are notoriously enigmatic and ambivalent.

In contrast to nineteenth-century English fiction, which is founded on "antagonisms" of class, manners, values, and sex, which are usually tempered, meliorated, and reconciled by marriage, renunciation, the maturation of the protagonist, or by an appeal to Christian or philanthropic values, Conrad's fiction, which charts the antagonisms sponsored by "the essential differences between the races" entertains but finally resists such conciliations. Hence, though Yanko manages to attach himself to a traditional biographical (fairy tale) "plot" when he rescues Swaffer's grandchild from "the sticky mud" and then subsequently marries Amy with a cottage for a dowry, his terrible end ("face down . . . his body in a puddle") simultaneously duplicates and reverses the moment that led this "most innocent of adventurers" to believe that he could reenter the human community. Similarly Amy's "act of impulsive pity" (she offered him "such bread as the rich eat in my country") by which "he was brought back again within the pale of human relations with his new surroundings" (121), and which transfigures her into "an angel of light," is also duplicated and reversed: her imaginative pity for and sexual attraction toward him yield to her "unreasonable terror, of that man she cannot understand." Hence when he is delirious and raves in Polish, Amy flees with the child in her arms unaware that she has rejected his raging cry "for water—only for a little water" (141). Amy's gift of bread and refusal of water justifies Kennedy's high claims for Yanko's fate as the most "*simply* tragic" he has ever encountered.

Moreover, though Yanko's last word "Merciful!" may seem the antithesis of Kurtz's last whispered cry "The horror! The horror!", yet it is also deeply ambiguous and profoundly disturbing. Yanko's cry may be a final testimony to Amy's "golden heart" (133); but he is after all delirious, and even several years after his death, Kennedy cannot be sure "she ever think[s] of the past" (142). Again Yanko's last word may be an appeal to the power of "his Redeemer" (129), so central to his life and culture, as it was to Conrad's own parents.[23] But as ever in his fiction the consolations of religion and even of self-knowledge are never fully available to his characters, and "the pale of human relations" in the domestic sphere is itself an extremely precarious and dangerous space. As we contemplate Kennedy's final estimation of Yanko's significance ("cast out mysteriously by the sea to perish in the supreme disaster of loneliness and despair"), our consolation resides in and is sponsored by the narrative's austere sympathy for all "the common people of the earth and the seashore" whether they dwell in the Carpathian mountains or in Colebrook. And, in marked contrast to Ford Madox Ford's rattling account of "the castaway" in *The Cinque Ports* (1900), which subsumes the reactions both of "the Marsh People" to the "German castaway" and of the "poor wretch" to his abusers into the omnivorous voice of the story-teller-as-gossip, Conrad's narrators assume and enact the burdens of transmission and interpretation imposed on them by the simplicity of those of "imperfect speech." Thus, finally, the great distinction of Conrad's finest narratives, such as "Amy Foster," is that they create within, and posit and solicit outside the tale, a community of listeners and readers who are persuaded to "collaborate with the author": to collaborate as he strives "to render the highest kind of justice" to the complex interaction generated by the "irreconcilable differences" of the "obscure minds" and "imperfect speech" of Yanko and the villagers, an interaction that ensures that Yanko struggles in vain to escape from "under the net."[24]

NOTES

All quotations from Conrad's works are taken from the Collected Edition (London: 1946); separate volumes are indicated by *CE* followed by the number.

1. Frederick R. Karl and Laurence Davies, eds., *The Collected Letters of Joseph Conrad*, vol. 2, 1898–1902 (Cambridge: 1986), 401–2. Henceforward *Letters*. Characteristically, Conrad's summary does not address the complex mediation of the tale. Clearly, however, for Conrad, Yanko, not Amy, is the center of the plot.
2. *Almayer's Folly*, 3. "Amy Foster" was first published in *Typhoon and Other Stories*, 1903, 105–42.
3. Z. Najder, *Joseph Conrad: A Chronicle* (Cambridge: 1983), 100–101. Compare Conrad's great letters to Garnett of June 19, 1896, and to Sir Henry Newbolt, June 5, 1905: "Other writers have some starting point. . . . They lean on dialect—or tradition—or on history—or on the prejudice or fad of the hour. . . . But at any rate they know something to begin with—while I don't" (*Letters* 1.288); and "Pray remember that from the nature of things I cannot count upon the moral support one's family, connections, the opinion of numerous early associates gives one against the hasty judgements of the world" (*Letters* 3.262).
4. *Letters* 2.348.
5. October 9, 1899, *Letters* 2.200.
6. Ian Watt, *Conrad in the Nineteenth Century* (Berkeley, CA: 1979), 44. Compare Daniel Schwarz's essay: "The reader recalls setting as a *paysage moralisé*, a mindscape engraved with the events that transpired" (see above, p. 175).
7. John Galsworthy, *Castles in Spain* (London: 1927), 78. This is a large issue but Conrad's British types, whether stolid like Captains McWhirr and Mitchell ("Typhoon" and *Nostromo*) or idealistic like Jim and Gould, are not formed abroad; the latter pair, despite their experiences in Patusan and Costaguana where they experience the differences between the races, do not feel profoundly their own strangeness. For Conrad neither intense curiosity nor alienation nor skepticism are associated with the Anglo-Saxon temperament.
8. In the holograph entitled "The Husband," dated June 18, 1901, both narrators are said to have suffered "the lot of the castaway." Hence as Kennedy remarks: "No matter how sure we may be of kindness we feel profoundly our own

strangeness . . . thrown suddenly by the sea upon the mercy of another race, whose tongue, thoughts, manners are a complete and momentous mystery. . . . Your difference creates a gulph—and there is no retreat" (24–25). It is difficult to resist the conclusion that Conrad cut this passage (as he later cut swathes from *Razumov*, the holograph of *Under Western Eyes*) because it might encourage irresponsible (and irritating!) speculation about the relationship between the creator and his creation. Cf. note 11.

I would like to thank the staff at the Beinecke Rare Book and Manuscript Library, Yale University, for their cooperation, and the curator, Vincent Girard, for his kind permission to quote from the manuscript.

9. Holograph, "The Husband," 5–6. This passage, not included in the published version, would have followed the frame-narrator's observation that Kennedy had "the gift of making people talk to him" (106). Perhaps Conrad cut the sequence because it was too pointed, even diagrammatic.

10. Conrad also played with the title "The Castaway" before he chose "Amy Foster" (June 13 and 19, 1901, *Letters* 2.332, 333). He did not settle on the present title until he made "certain corrections" to "the clean copy of The Castaway" in late mid-June.

Cf. notes 3 and 8. Conrad must surely have been aware of the obvious parallels between his own life and career and that of Yanko's. Thus briefly both share "not a little of that truly Polish hopefulness which nothing either nationally or individually has ever justified" (May 28, 1901, *Letters* 2.328); both share a "foreignness" that "had a peculiar and indelible stamp" (131–32); both were "adventurers at heart" who were "separated by an immense space from" their pasts "and by an immense ignorance from" their futures (*A Personal Record* 145); both fled Poland to escape "military service" (111). The ugly nickname "Yanko" (in Polish "little John") reminds us that Conrad could not abide distortions of "my real surname"—which (partly) led him to adopt a *nom de plume*.

11. Gustav Morf, *The Polish Heritage of Joseph Conrad* (New York: 1965 [first published 1930]), notes that Yanko's speech is saturated with expressions which the Western reader will take for an imitation of childish speech, but which is imitation Polish. Thus "steam-machines," rather than "engines," because of the Polish "maszyna"; and "iron track" from "Kolejzelazna" (216–17).

12. Juliet McLauchlan, "'Amy Foster'—Echoes from Conrad's Own Experience," *The Polish Review*, 23.3 (1978), 129.

13. M.M. Bakhtin, *The Dialogic Imagination*, ed. Michael Holquist, trans. Caryl Emerson and Holoquist (Austin, TX: 1981), 403.

14. This theme is central to Sanford Pinsker's essay on the tale, "'Amy Foster': A Reconsideration," *Conradiana* 9 (1977) 179–86.

15. Richard Herndon, "The Genesis of Conrad's 'Amy Foster,'" *Studies in Philology* 57: 549. Herndon sensibly reviews the biographical and psychoanalytical interpretations of the tale, and then proceeds to discuss its literary sources: namely Flaubert's "Un Coeur Simple," Dostoevsky's "The Idiots," and Ford's account of the German castaway in *The Cinque Ports* (Edinburgh: 1900), 162–63.

16. "A Familiar Preface" to *A Personal Record* (1912), xiii. Consider, for example,

Morf's rapturuous "vision" of the links between Conrad and Yanko via the theme of "a man who could not acclimatise himself in a foreign country," which concludes "have we not seen . . . the 'repressed' Conrad?" (*The Polish Heritage* 167–76).

17. Quoted by F. Berka, "Zhitomir and the Ukrainian Background in the Early Childhood of Joseph Conrad," *The Conradian* 8.2, 10.

18. See Apollo Korzeniowski's "Poland and Muscovy" and Bobrowski's *Memoirs*, quoted in *Conrad Under Familial Eyes*, ed. Z. Najder (Cambridge, Eng.: 1983) 75–88, 34. Andrej Busza, after a careful survey of Conrad's background, rightly concludes that these clashing sincere loyalties which he inherited and internalized "were representative of the political possibilities and modes of action open to an oppressed nation as a whole and gave him a great start as a writer" (Antemurale). Cf. also Norman Davies, *God's Playground: A History of Poland*, vol. 2 (New York: 1982), 29–60.

19. M.M. Bakhtin: *Problems of Dostoevsky's Poetics*, ed. and trans. Caryl Emerson (Minneapolis: 1984), 30, 27. "Thus," Bakhtin maintains, in a formulation that fits Conrad also, "the objective contradictions of the epoch did determine Dostoevsky's creative work—although not at the level of some personal surmounting of contradictions in the history of his own spirit, but rather at the level of an objective visualisation of contradictions as forces coexisting simultaneously (to be sure this vision was deepened by personal experience)."

20. Bakhtin, *The Dialogic Imagination*, 282, 291–92.

21. September 23–25, 1903, *Letters* 3.63; November 11, 1901, *Letters* 2.359. Cf. also Conrad's letter to Arnold Bennett, March 10, 1902, *Letters* 2.302.

22. *Letters* 2.200.

23. Cf. Conrad's letter to Garnett: "It's strange how I always, from the age of fourteen, disliked the Christian religion, its doctrines, ceremonies and festivals" (December 22, 1902, *Letters* 2.389–90).

24. March 22, 1902, *Letters* 2.394; Preface to *The Nigger of the "Narcissus."*

ENGLAND, ENGLISHNESS, AND THE OTHER IN E.M. FORSTER

Philip Dodd, University of Leicester

In E.M. Forster's *Where Angels Fear to Tread* (1905), the narrator generalizes the failure of the marriage of the Italian, Gino, and the Englishwoman, Lilia, in the following terms:

> No one realised that more than strong personalities were engaged; that the struggle was national; that generations of ancestors, good, bad, or indifferent, forbad the Latin man to be chivalrous to the northern woman, the northern woman to forgive the Latin man. (58)

The tone may usually be less solemn, and the address to the issue less direct, but the issue itself—the relationship between the English and others—is interrogated throughout Forster's works. One might say that the logic of the carefully structured volume of essays, *Abinger Harvest* (1936), which moves from England (the first essay is "Notes on the English Character") through immersion in "The Orient" (the title of the third section) and returns to England at the end, rehearses the general logic of Forster's work. It leaves England to return and "know the place for the first time." This essay will identify the terms in which Forster casts the relationship between England (emphatically not Great Britain) and non-English places, and to restore the intelligibility of these terms by explaining their provenance. The major focus of the study will be the novels, but reference will also be made to other Forster writings, on the understanding *not* that they are "background" to his fiction, but that *all* his writing is "creative," exploring the same nexus of issues.

Edward Said, in his brilliant work *Orientalism* (1978), has helped to recast our sense of the variety of ways dominant cultures shape and control other cultures subordinate to them. His argument has certainly contributed to my understanding of Forster, and not merely to his "Indian" books. For Said, the West has constituted a place called the "Orient," and it is his contention that

> without examining Orientalism as a discourse one cannot properly under-
> stand the enormously systematic discipline by which European culture
> was able to manage—and even produce—the Orient, politically, socio-
> logically, militarily, ideologically, scientifically and imaginatively during
> the post-Enlightenment period.[1]

The power of the West has been (in part) its power to represent the Orient in writing, and to fix its essence and that of the people:

> On the one hand, there was a collection of people living in the present; on
> the other hand, these people—as subject of study—became "the Egyp-
> tians," "the Muslims" or "the Orientals." . . . Every modern, native
> instance of behaviour became an effusion to be sent back to the original
> terminal, which was strengthened in the process. (234)

Now, clearly, Forster's concern with places as diverse as Italy (*Where Angels Fear to Tread, A Room With a View* [1908]), Egypt (*Pharos and Pharillon* [1923]), and India (*A Passage to India* [1924], *The Hill of Devi* [1953]) has hardly the systematic character Said ascribes to Orientalism. And yet following his argument, the places in Forster's fiction—including those within England—are never simply nor even primarily geographical locations, but sites of social and cultural value. Forster does not have to acquiesce in the settled valuations, but nor can he simply transcend them. To quote Said again:

> it must also be true that for a European or American studying the Orient
> there can be no disclaiming the main circumstances of *his* actuality: that he
> comes up against the Orient as a European or American first, as an
> individual second. And to be a European or American in such a situation
> is by no means an inert fact. It meant and means being aware, however
> dimly, that one belongs to a power with definite interests in the Orient,
> and more important, that one belongs to a part of the earth with a definite
> history of involvement in the Orient almost since the time of Homer. (11)

This argument makes us recognize that the relationship between England

and the Other (and not merely the Orient) is a relationship of power in Forster—one in which the English writer represents other places and peoples to the English.

Throughout his work Forster is careful not to represent other cultures to his English readers simply in exotic terms, and this is particularly so in comparison with some of his Imperialist contemporaries. This should hardly surprise us when we remember Forster's social location as a writer, a member of what Raymond William has called "The Bloomsbury Fraction," with its "hostility to imperialism, where the conscientious identification with victims was more negotiable than in England itself."[2] Consider, as an example of Forster's anti-exotic writing, "The Den," a short story first published in *The Nation and the Athenaeum* and later reprinted in *Pharos and Pharillon* (1923). The story is very brief, no more than nine-hundred words, in which the "I," an E.M. Forster persona, tells of his attempts to find an opium den in India and a hashish den in Egypt. The subject matter and locations lend themselves to exoticism, as a long tradition of such writing in English makes clear. Perhaps Thomas de Quincey's *Confession of an English Opium Eater* (1821) is the most famous example, with its opium dreams:

> Under the connecting feeling of Tropical heat and vertical sunlight, I brought together all creatures, birds, beasts, reptiles, all trees and plants . . . and assembled them together in China and Hindostan. From kindred feelings, I soon brought Egypt and her gods under the same law. I was stared at, hooted at, grinned at, chattered at, by monkeys . . . I fled from the wrath of Brama, through all the forests of Asia, Vishnu hated me.[3]

Forster's "The Den" uses the terms of the English exoticist tradition—there is the initiation of the "innocent" Englishman and the unknown, lurid places and peoples of the "Oriental" cities—but it does not ratify the tradition. The story begins with the speaker's announcement: "At last I have been to a den" (79). The reader's appetite for the customary revelations about the East is whetted. Yet the second sentence, beginning, "The attempt was first made many years ago" immediately suspends the promise of the first, and the reader is told the tale of the speaker's *failure* to see a den in Lahore. The desire for revelation is further frustrated as the speaker briefly rehearses his two subsequent failures to find hashish dens in Egypt. Only in the last paragraph is the promise of that opening sentence fulfilled. But it is realized in characteristic Forsterian fashion—the revelation is that there is no revelation:

And we found the genuine article at last. It was up a flight of stairs, down

207

which the odour (not a disagreeable one) floated. The proprietor—a one-eyed Maltese—battled with us at the top. . . . But we got in and saw the company. There is really nothing to say when one comes to the point. They were just smoking. And at the present moment they don't even smoke, for my one and only Den has been suppressed by the police. (81)

Forster both gives us hints of the monstrous "Orient" of the European imagination—the one-eyed Maltese is clearly a close relation of the Cyclops of Homer's *Odyssey*—but finally chastens and mocks that imagination. The cumulative disappointment that the narrator of "The Den" endures is the educative process of so many of Forster's protagonists when they travel abroad. Of course, the path to disenchantment is not readily taken by all of the characters. Consider, in *Where Angels Fear to Tread*, Philip's incapacity to abandon his vision of Italy and the Italians when he discovers that Gino's father is not an aristocrat but a dentist:

Philip gave a cry of personal disgust and pain. A dentist! A dentist at Monteriano. A dentist in fairyland! False teeth and laughing gas and the tilting chair at a place which knew the Etruscan League and the Pax Romana, and Alaric himself. . . . He thought of Lilia no longer. He was anxious for himself: he feared that Romance might die. (25)

Even the movement of a characteristic sentence in Forster seems to lead away from (false) expectation. Consider this sentence from *A Passage to India*: "They [Mrs Moore and Miss Quested] had made such a romantic voyage across the Mediterranean and through the sands of Egypt to the harbour of Bombay, to find only a gridiron of bungalows at the end of it" (26).

One might reasonably claim that Forster values his characters according to their willingness to be changed by what they encounter. For instance, at one pole in *Where Angels Fear to Tread* is Harriet: "And now Harriet was here acrid, indissoluble, large; the same in Italy as in England—changing her disposition never" (100). At the other pole is Caroline:

She had thought so much about this baby, of its welfare, its soul, its morals, its probable defects. But like most unmarried people, she had only thought of it as a word. . . . The real thing, lying asleep on a dirty rug, disconcerted her . . . now that she saw this baby . . . she had a great disposition . . . to exert no more influence than there may be in a kiss. (113)

What should happen to the English protagonists is perhaps most explicitly stated in *A Passage to India* when Miss Quested retracts her testimony against Aziz. "Although her hard school mistressy manner remained, she was no longer examining life, but being examined by it; she had become a real person" (238). Caroline in *Where Angels* recognizes the baby as a "real" thing; Miss Quested becomes a "real" person. Such characters travel to other cultures and, forced to confront the reality of the Other, they confront themselves. But clearly what is crucial in this process is the "reality" the English protagonists find in themselves. The non-English cultures and people are important primarily as catalysts to the English.

Given that this is the case, and bearing in mind Said's argument, it is impossible to understand what these non-English places and people signify in Forster's work without grasping what England and Englishness also signify. But before sketching what Englishness is in Forster's work, I want to enter a caveat about Said's thesis, which bears on the argument about Forster. Said's resolute and exemplary opposition to the appropriation of subordinated cultures by dominant ones leads him to insufficiently acknowledge that appropriation itself may be a recognition by the dominant culture of its own inadequacy. To phrase this in terms of Forster, the travels he and his characters make to other cultures are his clear recognition of the inadequacy of England and Englishness.

England is, not more than are Italy, Egypt, or India, simply a geographical space in Forster. Particularly during the early years of his life (he was born in 1879), England and Englishness were remade in dramatic ways. This is not the place to rehearse the general nature of such changes, which have been excavated, [4] but what is relevant here is that in the late nineteenth and early twentieth centuries the imaginative center of England and Englishness moved south, to London and the Home Counties. The shift from northern manufacturing industry to finance capital; the creation of a new elite, the civil services, and the professions; and the merging of the old industrial middle class into the commercial middle class—all these and other factors led to the transplantation to the south of the heart of England.[5] Recently reorganized educational institutions such as the public schools and Oxford and Cambridge became the cultural seedbeds of this "new" Englishness. As Matthew Arnold, one of its architects, put it in *Culture and Anarchy* (1869), to be a true Englishman one either had to belong to or be affiliated with these educational institutions. And as late as 1929, Bernard Darwin, in one of a large number of books written around that time about the public school system, said that, whatever one's views about it, "it is really to a great extent the English character that we are criticising."[6] To summarize briefly a complex argument, what was at the heart of this Englishness was a certain English masculinity taught in formal and informal ways, from the public schools where, according to one recent scholar, masculinity was at the core of the curriculum, to style manuals that identified certain ways of writing as properly masculine and English.[7] Reviews of literature would often formulate praise in national terms. For instance, in *Quarterly Review* the poet Alfred Austin was praised as a good example of a "concrete individual Englishman," for he wrote poetry of "healthy directness."[8]

Clearly, in terms of his biography, Forster was "English." Born in London, he was sent as a day boy to Tonbridge Public School from 1893-97, and moved on to Cambridge where he first read classics and then history. But Forster's affiliation to Englishness was more than a biographical matter. In his writings, whatever his judgment about the value of the Englishness forged in these new social locations, he never seriously challenged the view that these were the seats of Englishness. Even in his script for Humphrey Jennings's Second World War documentary, *A Diary for Timothy* (1944), Forster assumes that southern landscape is the heart of England and the film moves outward from that heart. To see his understanding of Englishness in general terms, one only has to read "Notes on the English Character," where he identifies the English as middle-class and their formative institutions as the "Public Schools"[9] or the novel *Maurice* (published posthumously in 1971, written in 1913–14), with its choice of a hero who passes through the institutions of the Public School, Cambridge, and the City. The narrator describes Maurice's fate after Cambridge thus: "Maurice was stepping into the niche that England had prepared for him" (53).[10] Forster's English locations are always those of southern rural England. Note the sentence from *Howard's End* (1910): "The station, like the scenery, struck an indeterminate note. Into which country will it lead, England or suburbia?" (16).

But what is interesting in Forster's fiction is that England and Englishness are not simply the framework of the books, but are themselves subject to interrogation. Consider, for example, Maurice's disturbed recognition that his homosexual nature challenges his received English masculinity. Maurice puts it this way to Durham, who has approached him with protestations of desire: "Durham, you're an Englishman, I'm another. Don't talk nonsense" (57). Or consider, in *The Longest Journey* (1907), how central to the book is a criticism of the English masculinity that Sawston School produces: "it aimed at producing the average Englishman, and, to a very great extent, it succeeded" (48).

One might speculate that what allowed Forster, this quintessential Englishman, to stand at least partly outside this Englishness was his homosexuality. To use the phrase he himself employed to describe another homosexual, the Greek poet C.P. Cavafy, Forster stood at a "slight angle" to Englishness.[11] And since nowhere in the ideological landscape that was England could he find a place for his own nature, he went elsewhere to find places where that nature might be expressed. At least, whatever the accuracy of this speculation, it is certainly the case that England and the English are throughout Forster's writing opposed to "nature." Englishmen and "nature" are mutually incompatible terms. Consider, for example, Mrs Herriton in *Where Angels Fear to Tread*, who sends Lilia, her daughter-in-law, to Italy chaperoned by Miss Abbott. It is made perfectly plain that Lilia is "subject" at the beginning of the book to Mrs Herriton in all things (10). When the letter arrives from Yorkshire telling Mrs Herriton that Lilia is engaged to Gino, it is entirely appropriate that Mrs Herriton should be in the garden sowing some vegetables. She is a "grower" of all things, human and natural. Interrupted and disturbed by the news the letter brings, Mrs Herriton returns to the garden only at the end of the day:

> Just as she was going upstairs she remembered that she never covered up those peas. . . . Late as it was, she got a lantern from the toolshed and went down the garden to rake the earth over them. The sparrows had taken every one. But countless fragments of the letter remained, disfiguring the tidy ground. (19)

The implications of the scene could hardly be clearer—the attempt to protect her garden or Lilia from nature is equally doomed. We can see this even at the end of the novel. Philip and Caroline on the train, after the latter's declaration of her love for Gino, "hurried back to the carriage to close the windows lest the smuts should get into Harriet's eyes" (160). Heroines from Jane Eyre through Dorothea Brooke to Ursula Brangwen may have stared out of windows, yearning to be united with the world, but Harriet must have the window closed on her.[12] The English in Forster's works are proud of their determination to repress, or at least control, nature. Not only is there Mrs Herriton's "tidy ground," but the "arid tidiness" of the civil lines in *A Passage to India* (1924). And there is Mrs Failing's celebration of conventions (rather than nature) in *The Longest Journey*:

> I say, once more, beware of the earth. We are conventional people, and conventions—if you will but see it—are majestic in their way, and will claim us in the end. We do not live for great passions or for great memories. (274)

In case we should be led to believe, despite the import of the last quotation, that nature in Forster is primarily the nonhuman world, we only have to remember Forster's novel *Maurice*, which, since he did not intend to publish it, is more explicit on most matters than his other writing. There is the description of Maurice's clothes as a "carapace" and the explicit declaration of the doctor in the face of Maurice's questions about the likelihood of changes in the law in England about homosexuality: "England has always been disinclined to accept human nature" (185).

But my implicit claim that England represses "nature" and that in other countries it is acknowledged is an inadequate formulation as it stands. If Forster does not see other cultures in terms of exoticism, he does not simply view them as "nature." For instance, throughout *Where Angels Fear to Tread* Forster is determined to claim that Italian life has its own culture—and not one that in any absolute way is preferable to that of England:

> Italy is such a delightful place to live in if you happen to be a man. There one may enjoy that exquisite luxury of Socialism—that true Socialism which is based not on equality of income or character, but on the equality of manners . . . the brotherhood of man is a reality. But it is accomplished at the expense of the sisterhood of women. (42)

And in *A Passage to India*, one would hardly accuse Forster of trying to avoid some of the complexities of Indian cultures. Forster is at pains to register the specific culture that nurtured Aziz:

> He had always liked this mosque. It was gracious, and the arrangement pleased him. . . . The front—in full moonlight—had the appearance of marble and the ninety-nine names of God on the frieze stood out black. . . . Here was Islam, his own country, more than a Faith . . . Islam, an attitude towards life, both exquisite and desirable, where his body and his thoughts found their home. (20)

Rather than in the cultures to which Forster's protagonists travel, "nature" is focused in one male character encountered in non-English cultures, for example Gino in *Where Angels Fear to Tread*, or George in *A Room with a View*, or Aziz or the punkah Wallah in *A Passage to India*. Consider these brief passages from a number of Forster's works:

> For a young man his face was rugged, and—until the shadows fell upon it—hard. Enshadowed, it sprang into tenderness. She saw him once again at Rome, on the ceiling of the Sistine Chapel, carrying a burden of acorns. (*A Room with a View* 30)

> Almost naked and splendidly formed . . . he caught her attention as she came in. . . . This man would have been noticeable anywhere; among the thin-hammed, flat-chested mediocrities of Chandrapore he stood out as divine. (*A Passage to India* 212)

You're taking me wrongly. I'm in love with Gino—don't pass it off—I mean it crudely—you know what I mean. . . . He's never flattered me nor honoured me. But because he's handsome, that's been enough. (*Where Angels Fear to Tread* 158)

In *Erewhon* (1872), on which Forster wrote under the title, "A Book That Influenced Me," Samuel Butler wrote of a society built upon the principle that "the body is everything."[13] That is too crude a formulation of Forster's position in his own writing, but his "journeys" in his fiction and nonfiction may be seen as ones in search of a place where the English will be forced to confront the "nature" that England and Englishness have repressed. It is worthwhile noting that the earlier quotation from *A Passage to India* (20) claimed that Islam offered Aziz a home for his body and his thoughts— precisely what England will not offer the English. And to emphasize that England is less a geographical than an ideological space, one has only to note that in *The Longest Journey* and *Maurice*, both of which are set exclusively in England, Forster is able to find persons, Robert and Alec respectively, who stand outside "England," live more fully in the body, and become objects of erotic concern. That there is correspondence in Forster between travels across cultures *within* England and travels outside England can be found in the fact that the "erotic" figures from the English and non-English places are lower in social class than the English protagonists whom they awaken. This is as clear with Alec (*Maurice*) or with Robert (*The Longest Journey*) as it is with Gino (*Where Angels*) or Aziz (*A Passage*). Raymond Williams's comment quoted earlier may help us to understand why Forster found it easier to journey outside England, rather than within it, to find a place from which to mount a cultural critique of England—it was, as Williams says, easier to affiliate with victims elsewhere than in England, given the relative threat of each to the ideological England to which Forster belonged.

It should be clear now what I believe the relationship between the English and others to be in Forster—a way in which he can interrogate his Englishness, identify what it represses, and allow that repressed nature expression, even if not necessarily fulfillment (consider Caroline at the end of *Where Angels*). Elsewhere is a critique of Englishness. And yet to say this is to tell only part of the story—for what has been said can seem to imply that the "resolution" in Forster's fiction is primarily a personal matter, rather than, as I now wish to explore, a matter of more public import. The relationship of an English man or woman with a non-English male is offered as a utopian glimpse of the possible relationship between cultures—dominant and subordinated ones. To give substance to this assertion, I need to sketch the

213

cultural resonances of "nature," and particularly of male friendships as the incarnation of nature in the early twentieth century.

John Alcorn's *The Nature Novel from Hardy to Lawrence* (1977), a study of the "rediscovery" of nature by writers around the turn of the century, is a useful starting point. This is Alcorn's definition of a naturist writer—one to which my account of Forster in part answers:

The naturist is a child of Darwin; he sees man as part of an animal continuum; he reasserts the importance of instinct as a key to human happiness; he tends to be suspicious of the life of the mind; he is wary of abstractions. He is in revolt against Christian dogma, against conventional morality, against the ethic which reigns in a commercial society.[14]

This absorption, of course, takes in more than novelists—in order to suggest how common among a certain group of writers is the stance that Forster adopts toward England, let me briefly compare Forster's terms of reference with those of an apparently dissimilar writer, Edmund Gosse, in his autobiography, *Father and Son*, published in 1907, the same year as the *The Longest Journey*. Philip Gosse's "trim and polished garden" is not unlike that of Mrs Herriton's in *Where Angels Fear to Tread*.[15] The question asked in *A Room with a View*— "Do you suppose there's any difference between spring in nature and spring in man?" (71)—might well be the question the son poses to the father in Edmund Gosse's book. And it is entirely predictable that not only Forster, but Gosse too, should oppose pagan Italy to Protestant England: Gosse builds into *Father and Son* the information that he writes the book by the Arno, the river of Florence. As Sheila Rowbotham says in *Socialism and the New Life: The Personal and Sexual Polices of Edward Carpenter and Havelock Ellis* (1977), "Italy was the symbolic alternative to late Victorian middle-class values."[16] Although this is clearly the case, one needs to add that Forster's Italy is different from that of, say, Walter Pater, whose representative is the aesthete Cecil in *A Room with a View*.

If this establishes the commonness of the currency Forster used, to grasp what Forster contributes to this currency we need to understand how he draws upon an important but still neglected strain of writing and thought of the period—at the center of which is Edward Carpenter. Carpenter himself, on whom Forster wrote an essay, is memorably described by Forster in his "Terminal Note" to *Maurice*:

He was a socialist who ignored industrialism and a simple-lifer with an independent income and a Whitmanesque poet whose nobility exceeded his strength, and finally, he was a believer in the love of comrades, whom he sometimes called Uranians. . . . For a short time he seemed to hold the key to every trouble. (217)

The language of Carpenter is evident in Forster's work. Consider how this statement from Carpenter echoes matter already quoted from Forster: the starving of the human heart, the denial of the human body and its needs, the huddling concealment of the body in clothes.[17] In *Towards Democracy* (1883), he dramatizes the upper-middle classes living obediently in "the prison life of custom without the touch of nature."[18]

Perhaps predictably, Forster's most explicit homage to Carpenter can be found in *Maurice* where the protagonist uses Carpenter's own term "comrade" to describe Alec and himself:

Maurice took [his hand], and they knew at that moment the greatest triumph ordinary man can win. Physical love means reaction, being panic in essence, and Maurice saw how natural it was that their primitive abandonment at Penge should have led to peril. . . . And he rejoiced because he had understood Alec's infancy through his own, glimpsing not for the first time, the genius who hides in man's tormented soul. Not as a hero, but as a comrade, had he stood up to the bluster. (198)

It is in the implications of the word "comrade"— used explicitly only in *Maurice*, but implicitly throughout Forster—that we can understand how the expression of what may seem a private matter has public and even political reverberations. The acknowledgment by Forster's English characters of their intimacy with the men from subordinated cultures, the embrace of them as "comrades" offers not only a glimpse of the possible relationship between individuals, but also between cultures. "Comrade" has not only the homosexual connotation already noted—the *OED* gives an early meaning as "sharing a bedroom"—but also a socialist or at least democratic one. The most explicit acknowledgment of the possible reverberations of the acknowledgment of intimacy with the Other can be found in *A Passage to India*. Of Ronny, the narrator says, after his outburst against the Indians: "One touch of regret—not the canny substitute but the true regret from the heart—would have made him a different man, and the British Empire a different institution" (50).

If explicit here, it is implicitly present in much of Forster. Note the description of Lucy near the end of *A Room with a View*: "He gave her a sense of deities reconciled, a feeling that, *in gaining the man she loved, she would gain something for the whole world*. . . . He had robbed the body of its taint, the world's taints of their sting; he had shown her the holiness of direct desire" (my emphasis, 218). For Forster's simultaneous resolution of personal needs and cultural ones in the discovery of a male from a culture outside England-as-an-ideological-space, one can find a mirror in *The Memoirs of John Addington Symonds* (1984), the long-suppressed memoirs of Symonds's life as a homosexual. In that book, Symonds describes his relationship with a Swiss man from a lower-class family, and says:

> [I was] struck with the wonder of discovering anyone so new to my experience, so dignified, so courteous, so comradely, realizing at one and the same time for me all that I had dreamed of *the democratic ideal and all that I desired in radiant manhood.* [19] (My emphasis.)

It is important to add, though, that what separates Carpenter from Forster and Symonds is that he had a politics—socialism—that offered to hold together the public and private spheres. In Forster (as well as Symonds), there is no sense of how this utopian bonding between individuals can be generalized.

In Forster, then, I have tried to show that non-English cultures and people are not simply reduced to an exotic status, nor are they objects of aesthetic contemplation, nor even do they in any simple terms play Nature to England's Culture. Rather what the English—or rather the best English—discover on their travels to other cultures is an intimacy with the Other, often from a "lower" or subordinated culture, an intimacy that stands as a utopian model of what might come to pass not only between individuals but also between cultures. Said's thesis holds—Forster does bestow meaning on other places and people from a position of power and contains concern with those cultures within an Anglocentric framework. But in doing so he acknowledges the inadequacy of the Englishness that he and his protagonists take on their travels with them. When William Walsh, one of the founding fathers of Commonwealth literature, of which Forster's *Passage to India* is a canonical text, said in his book *Commonwealth Literature* (1973) that Indians do not write in a "direct masculine way" but with "Indian tenderness," he was, without knowing it, doing what Forster had done before him—finding in cultures other than the English what the English themselves repressed.[20]

NOTES

Quotations from Forster's works are from the following editions: *Where Angels Fear to Tread* (Harmondsworth: 1959); *Pharos and Pharillon* (London: 1923); *Abinger Harvest* (London: 1936); *Maurice* (Harmondsworth: 1972); *Howards End* (Harmondsworth: 1975); *The Longest Journey* (Harmondsworth: 1974); *A Passage to India* (Harmondsworth: 1961); *Two Cheers for Democracy* (London: 1951).

1. Edward Said, *Orientalism* (London and Henley: 1978), 3.
2. Raymond Williams, "The Bloomsbury Fraction," *Problems in Materialism and Culture* (London: 1980), 156.
3. David Masson, ed., *The Collected Writings of Thomas de Quincey* (Edinburgh: 1896–97), vol. 3, 442. For a fine account of de Quincey's "Orientalism," see Robert M. Maniquis, "Lonely Empires: Personal and Public Vision of Thomas de Quincey," *Literary Monographs*, 8, ed. Eric Rothstein and Joseph Anthony Wittreich Jr., (Madison: 1976), 49–127.
4. See Robert Colls and Philip Dodd, eds., *Englishness: Politics and Culture 1880–1920* (London: 1986).
5. Alun Howkins, "The Discovery of Rural England," *Englishness*, 62–88.
6. Matthew Arnold, *Culture and Anarchy, The Complete Prose Works of Matthew Arnold*, ed. R.H. Super (Ann Arbor, MI: 1965), vol. 5, Preface; Bernard Darwin, *The English School* (1929) quoted in Martin J. Weiner, *English Culture and the Decline of the Industrial Spirit* (Cambridge, Eng.: 1981), 21.
7. On masculinity, see J.A. Mangan, *Athleticism in the Victorian and Edwardian Public School: The Emergence and Consolidation of an Education Ideology* (Cambridge, Eng.: 1981), 135, and Quiller Couch, *On the Art of Writing* (1916). For explorations of the construction of masculinity during this period, see Philip Dodd, "Englishness and the National Culture," *Englishness*, 1–28.
8. Quoted in C.K. Stead, *The New Poetic* (London: 1967), 75.
9. In *Abinger Harvest*, 3–14.
10. Though first published in 1971, *Maurice* was written during 1913–14.
11. "The Poetry of C.P. Cavafy,"*Pharos and Pharillon*, 91.
12. The references are to Charlotte Brontë's *Jane Eyre*, George Eliot's *Middlemarch*, and D.H. Lawrence's *The Rainbow*.
13. In *Two Cheers for Democracy*, 224–28.
14. John Alcorn, *The Nature Novel from Hardy to Lawrence* (London: Macmillan, 1977), x.
15. Edmund Gosse, *Father and Son*, ed. James Hepburn (London: 1974), 135.
16. Sheila Rowbotham and Jeffrey Weeks, *Socialism and the New Life: The Personal and Sexual Politics of Edward Carpenter and Havelock Ellis* (London: 1977), 31.
17. *My Days and Dreams* (1916), quoted in *Socialism and the New Life*, 27.
18. *Towards Democracy* (London: 1883), 31.
19. *The Memoirs of John Addington Symonds*, ed. and intro. Phyllis Grosskurth (London: 1984), 263.
20. William Walsh, *Commonwealth Literature* (Oxford: 1973), 1, 10.

NEW WORLDS FOR OLD: H.G. WELLS

Patrick Parrinder, University of Reading

"I am English by origin," wrote H.G. Wells in the 1930s, "but I am an early World-Man and I live in exile from the world community of my desires." [1] All three parts of this statement—Wells's English origins, his internationalist outlook, and his sense of exile from a longed-for new world—deserve to be emphasized, though it is his militant sense of world citizenship that most obviously sets him off from other writers of his time. A founder of P.E.N. (Poets, Essayists, Novelists), he was elected international president of that organization in 1934. In the same year, some of his admirers banded together to form the first H.G. Wells Society. They debated whether to change its name to the Open Conspiracy (after Wells's book of 1928 advocating a popular movement for world government), but in the end they decided to call it Cosmopolis.[2] The Cosmopolitans, like thousands and perhaps millions of Wells's readers, had been inspired by the gospel of human unity that he had expounded in primers such as *The Outline of History* (1920) and *The Work, Wealth and Happiness of Mankind* (1931), in pamphlets and newspaper articles, and in novels and stories since the end of World War I.

By 1934, as a result of well-publicized meetings with Franklin D. Roosevelt and Joseph Stalin, Wells had briefly become a kind of unofficial world statesman. His project of high-minded conversations with world leaders was soon overtaken by events, but his son, Anthony West, has claimed that between the wars he did "as much as any man then living to create the climate of opinion in the middle ground that was to make the creation of the United Nations and the establishment of the European Economic Community . . . inevitable."[3] Many others have seen Wells as a failed prophet of world order, however. For some, his visions of global and cosmic integration in such works as the future-history novel *The Shape of Things to Come* (1933) and in its movie version *Things to Come* (1935) can now be relegated to the storehouse of discredited collectivist fantasies.[4] But

219

Wells is a complex writer whose literary career spanned five decades and who continues to command attention largely as a result of the imaginative intensity of his early work—the sequence of classic science fiction novels, the social comedies, and (to a much lesser extent) the novels of ideas that he wrote before 1914.

Though he may be called the prototype of the modern politically conscious international writer, Wells's cosmopolitanism emerged from the peculiar conditions of the late nineteenth century. Describing himself as an "early World-Man," he invoked the Darwinian outlook that had been impressed upon him by his great biology teacher T.H. Huxley. *Homo Sapiens*, the World-Man, was, he believed, struggling to evolve out of the divided humanity of the era of sovereign nation-states. Only by evolving into World-Men could humanity survive the modern industrial age in which, for the first time, the species was capable of bringing about its own extinction. Wells's ideal of world government was first conceived at the height of the age of European Imperialism, and in some respects he preached a kind of super-Imperialism that remained rooted in the ideology of Empire. Though an outspoken critic of conventional Imperialism, he looked to a new civilization unified in its attachment to Western rationality, with a centralized government run by a scientific elite and combining moral authority with military strength.

The years 1880–1920, during which nearly all of Wells's major works were written, now appear not only as the heyday of competitive world-conquest but as the period of a resurgence of cultural nationalism in England itself.[5] Wells had little or no political sympathy with the advocates of "Englishness," but his English origins are unmistakable and his early writing did much to fix a certain image of ordinary English life in his reader's minds. It is, therefore, necessary to describe Wells's Englishness in order to appreciate his wider outlook. He may be classed as a "provincial" writer in more than one sense of the term. No doubt there is something to be said for his attempts to construct a religion of humanity without Comte and a socialism without Marx, but his hostility to these two thinkers partly reflects his sense that their methodical, system-building habits were alien to the English mind.[6] On a more positive note, he is provincial in that he grew up on the underside of the British class system—he was the son of a small shopkeeper and a lady's maid—and his early life was spent in the small towns and villages of South-East England. He was not a Londoner by birth or upbringing, and it was thanks to the newly instituted scheme of government scholarships for needy students that, at the age of eighteen, he moved to the metropolis. Wells was later to re-create the scenes of his Home Counties childhood with warmth and nostalgia in novels such as *The Invisible Man* (1897) and *The History of Mr Polly* (1910), written when the new rural ideal of Southern English life was emerging.

No sooner has he imagined a rural idyll, however, than Wells sets out to disrupt it—as both *The Invisible Man* and *Mr Polly* bear witness. Wells's own father, we are told, "grew up to gardening and cricket [he played for Kent], and remained an out-of-doors, open-air man to the day of his death"—and what could be more English than that?[7] Yet Wells seems to have inherited his restlessness from his father who (as his son recalled in 1906) "still possesses the stout oak box he had made•to emigrate withal, everything was arranged that would have got me and my brothers born across the ocean, and only the coincidence of a business opportunity and an illness of my mother's, arrested that."[8] Joseph Wells had tried and failed to break free from the routine of life in the Home Counties. His son was luckier and more successful. As his narrator George Ponderevo says in *Tono-Bungay* (1909), "One gets hit by some unusual transverse force, one is jerked out of one's stratum and lives crosswise for the rest of the time, and, as it were, in a succession of samples. That has been my lot."[9] George reaches the heights of London society as a result of his uncle's runaway success with a patent-medicine business, but at the end of the novel, like Joseph Wells, he seems to be looking away from Europe toward the New World. He has become a naval architect whose pet project (a destroyer) "isn't intended for the Empire, or indeed for the hands of any European power." "I have come to see myself from the outside, my country from the outside—without illusions," George adds.[10] This was certainly Wells's aim.

I shall argue that, metaphorically and to some extent literally, the New World was the necessary foil to the bankrupt Old World in Wells's writing. In a 1915 study of *The World of H.G. Wells*, which was much the best book written on its subject during his lifetime, Van Wyck Brooks argued that there was a natural affinity between Wells and his American readers. "His mind" wrote Brooks, "is a disinherited mind, not connected with tradition, thinking, and acting *de novo* because there is nothing to prevent it from doing so."[11] Wells's sense of disinheritance may doubtless be attributed to his insecure childhood (the family broke up when he was thirteen) and to the disastrous physical injury that brought him near to death as a young adult; nevertheless, such experiences are common enough. One frequent recourse of the disinherited is an aggressive, chauvinistic identification with the aims of a particular group of society, and this is a prime cause of modern nationalism; but Wells's identification, instead, was with the global aims of modern socialism and modern science. It was likely, also, as a "disinherited" writer, that he would veer between autobiographical fiction and outright fantasy; his was too restless an imagination to stay for long in secure possession of a particular social world, as Jane Austen and Anthony Trollope had done. In later life Wells took such opportunities as came his way to travel the world, and for a time he maintained a house in France. These experiences were reflected in the widening horizons of his autobiographical novels, but they only sharpened his sense of the global need for reconstruction and change. He remained in exile from the community of his desires.

In both fictional and nonfictional forms, Wellsian autobiography follows the conventional shape of the *Bildungsroman*, in which the hero progresses from narrow origins to a position reflecting the author's general view of human life. Wells's heroes may be divided into those who share his authorial consciousness (reflected in the title of the triumphal final chapter of his *Experiment in Autobiography* [1934], "The Idea of a Planned World"), and those who do not. The latter—comic heroes such as Kipps, Bert Smallways, and Mr Polly—remain limited, provincial, and English. Art Kipps, an orphan growing up in New Romney in Kent, finds his first home in the backyard and kitchen of his uncle's shop in the High Street, and especially in the corner under the ironing-board where, with the aid of an old shawl, he makes a cubby-house. This "served him for several years as the indisputable hub of the world."[12] The wider horizons opened up for him as a young man by a timely legacy leave no lasting impression, and finally we see the mature Kipps happily ensconced in another small High-Street shop, this time in Hythe (which is only a few miles from New Romney). Alfred Polly, too, begins in obscurity and ends in a similar though more contented state at the Potwell Inn, a deeply rural (and highly unlikely) English Rabelaisian paradise. Bert Smallways, in the scientific adventure-story *The War in the Air* (1908), gets carried off from Dymchurch beach on the Kent coast in a balloon, and is then involuntarily transported across the Atlantic with the Kaiser's Zeppelin fleet. Bert witnesses the aerial siege of New York and a full-dress battle at Niagara Falls between German and Japanese aviators. He falls into possession of the plans for a new flying-machine, which he is able to personally hand over to the president of the United States. But Bert is still a Kentish shop-boy at heart, and the president sends him back to England, where he is reunited with his Edna and settles down in his home town of Bun Hill; meanwhile, industrial civilization is destroyed by war and plague. In *The Outline of History*, Wells famously described human life as a "race between education and catastrophe,"[13] and in Kipps, Polly, and Smallways he outlined the comedy of stubbornly uneducated lives, not World-Men but little Englishmen, who could take no part in that race.[14]

Kipps and Polly had been sent to dingy private schools. In *The History of Mr Polly*, Wells uses a memorable image to indicate what a proper education might have done for his hero:

I remember seeing a picture of Education—in some place. I think it was Education, but quite conceivably it represented the Empire teaching her

Sons, and I have a strong impression that it was a wall-painting upon some public building in Manchester or Birmingham or Glasgow, but very possibly I am mistaken about that. It represented a glorious woman, with a wise and fearless face, stooping over her children, and pointing them to far horizons. The sky displayed the pearly warmth of a summer dawn, and all the painting was marvellously bright as if with the youth and hope of the delicately beautiful children in the foreground. She was telling them, one felt, of the great prospect of life that opened before them, of the splendours of sea and mountain they might travel and see, the joys of skill they might acquire, of effort and the pride of effort, and the devotions and nobilities it was theirs to achieve. Perhaps even she whispered of the warm triumphant mystery of love that comes at last to those who have patience and unblemished hearts. . . . She was reminding them of their great heritage as English children, rulers of more than one-fifth of mankind, of the obligation to do and be the best that such a pride of empire entails, of their essential nobility and knighthood, and of the restraints and charities and disciplined strength that is becoming in knights and rulers . . .
  The education of Mr Polly did not follow this picture very closely.[15]

The picture of education belongs in the bombastic tradition of Victorian political allegory. Wells's mockery distances him from the representation of the "pride of empire," though the passage certainly draws on the romantic allure of Imperialist mythology. Characteristically, Wells sets out to channel such feelings in the direction of cosmopolitan idealism. In *The Outline of History*, there is an illustration by J.F. Horrabin showing Britannia, Germania, Marianne, and other "Tribal Gods—*natural symbols for which men would die*—of the Nineteenth Century," Nevertheless, in the *Autobiography* Wells attributes his childish sexual awakening to his "naive direct admiration for the lovely bodies, as they seemed, of those political divinities of Tenniel's in Punch, and . . . the plaster casts of Greek statuary that adorned the Crystal Palace."[16] The picture of Education stands somewhere between these figures and the representation, in Wells's modern Utopia, of a Utopian coin portraying not Britannia, as the old English penny piece did, but "Peace, as a beautiful woman, reading with a child out of a great book, and behind them are stars, and an hour-glass, halfway run."[17] In *Mr Polly*, Imperial Education is beckoning to the horizon at dawn rather than reading to her children at twilight, suggesting a stirring of energies no longer present in Utopia.
  Just as the picture of Peace excludes the adult male (normally associated with violence), the picture of Imperial Education offers no position for the disinherited. In that respect, it is simply irrelevant to the needs of *Mr Polly*, and remains no more than a dream for Wells himself. He could not aspire to

"nobility and knighthood," nor could he march toward the "great prospect of life" in a straight line. Instead, disentanglement from the confined world of his upbringing came on Wells very suddenly. Van Wyck Brooks, no doubt influenced by George Ponderevo's reflections in *Tono-Bungay*, expressed this as follows:

> The world of shopkeeping in England is a world girt about with immemorial subjections; it is, one might say, a moss-covered world; and to shake oneself loose from it is to become a rolling stone, a drifting and unsettled, a detached and acutely personal, individual. It is to pass from a certain confined social maturity, a confused mellowness, into a world wholly adventurous and critical, into a freedom which achieves itself at the expense of solidity and warmth.[18]

It is profoundly significant, in this light, that Wells launched his literary career not with autobiographical fiction (this was to come slightly later) but by leaping with one bound from the familiar to the exotic, from the Home Counties to the "wholly adventurous and critical" world of his short stories and scientific romances.

In many of the early stories the theme of escape is paramount, as the hero undergoes an "out of body" experience (extending, in "Under the Knife" [1897] to a cosmic journey to the other end of the galaxy), or disrupts the frame of experience in some way (as in "The Man Who Could Work Miracles" [1899], where Mr Fotheringay manages to stop the earth's rotation). The geography of Imperialism is reflected in stories such as "In the Avu Observatory" (1895) set in Borneo; "The Treasure in the Forest" (1895) and *The Island of Doctor Moreau* (1896) in the South Pacific; "Aepyornis Island" (1895) in the Carribean; "The Empire of the Ants" (1911) and "The Country of the Blind" (1911) in South America; and "The Pearl of Love" in ancient India—not to mention "In the Abyss" (1897) on the ocean bed and "The Crystal Egg" (1899) on Mars. In other stories the homely English setting is disrupted by strange events, which are often exotic in the strict sense. "New Genus, by heavens! And in England!"[19] exclaims the entomologist Haply, confronted by the phantom of a strange moth in an airtight laboratory somewhere in Kent ("The Moth" [1895]), while "The Flowering of the Strange Orchid" (1895) takes place in a Home Counties greenhouse. Davidson, in "The Remarkable Case of Davidson's Eyes" (1895) simultaneously experiences life in London and on Antipodes Island.

In the Preface that he wrote for *The Country of the Blind and Other Stories* (1911), Wells looked back on the spontaneous and, as it were, irresponsible conception of his early stories:

I found that, taking almost anything as a starting-point and letting my thoughts play about it, there would presently come out of the darkness, in a manner quite inexplicable, some absurd or vivid little incident more or less relevant to that initial nucleus. Little men in canoes upon sunlit oceans would come floating out of nothingness, incubating the eggs of prehistoric monsters unawares; violent conflicts would break out amidst the flower-beds of suburban gardens; I would discover I was peering into remote and mysterious worlds ruled by an order logical indeed but other than our common sanity.[20]

This passage is rich in Wellsian motifs. The apparent interchangeability of the settings is very notable, though it may be added that, where the "little men in canoes" were a staple of late Victorian adventure fiction, the violent conflicts in suburban gardens are typical of Wells's scientific romances. In *The War of the Worlds* (1898), the first Martians land at Woking in Surrey, where Wells was then living. The initial idea for the story came from his brother Frank. Wells, who had just learned to ride a bicycle, "wheeled about the district marking down suitable places and people for destruction by my Martians."[21] Today the sandpits on Horsell Common near Woking are still instantly recognizable to a reader of *The War of the Worlds*. Nevertheless, in the many media adaptations of the book its original setting has almost invariably been discarded; the local realism of the story, which works out triumphantly on the printed page, turns out to be interchangeable after all. In Orson Welles's 1938 adaptation for CBS radio the Martians land in New Jersey, while in George Pal's 1953 movie version they attack Northern California. A mass panic comparable in scale to that aroused by the Orson Welles dramatization was reported from Ecuador, where in 1949 a "localized version" of *The War of the Worlds* broadcast in Quito led to a riot in which the crowd stormed and set fire to the radio station.[22] Wells's tale had proved to be universal even if his chosen setting was highly particular.

The Wellsian scientific romance combines irresponsible imagination with the disciplined working-out of the initial hypothesis. To that extent, it reflects the scientific ideal and may even run parallel to the processes of scientific explanation. Its cosmic outlook is based in late nineteenth-century physics and astronomy and, above all, in the vistas of earth history opened up by the discovery of evolution and the geological record. When Wells dreamed of "peering into remote and mysterious worlds" or of "incubating the eggs of prehistoric monsters," he was turning his scientific studies to imaginative account. The notion of "peering into" strange worlds irresistibly suggests the telescope or microscope. In later years he would claim that the "central fact" of his student years spent in the South Kensington laboratories and dissecting rooms, and in the galleries of the new Natural

History Museum, was his attendance at Huxley's course in comparative anatomy. With that as a foundation he acquired "what I still think to be a fairly clear, and complete and ordered view of the ostensibly real universe . . . I had man definitely placed in the great scheme of space and time."[23] Such a movement from the particular to the universal remained essential to his notion of disciplined imagination, as is clear from a famous exchange with Joseph Conrad which he records in his *Autobiography*. Lying on Sandgate beach, the two men debated the best way to describe a boat that they could see riding out in the water:

> it was all against Conrad's over-sensitized receptivity that a boat could ever be just a boat. He wanted to see it and to see it only in relation to something else—a story, a thesis. And I suppose if I had been pressed about it I would have betrayed a disposition to link that story or thesis to something still more extensive and that to something still more extensive and so ultimately to link it up to my philosophy and my world outlook.[24]

For Wells, in other words, it is not just a boat but a specimen, a model—as likely as not, a model of social experience and social relations. Both his science fiction and his social fiction rely on different kinds of model-building.

In science fiction, Wells's most influential model was doubtless the "Man of the Year Million," originally proposed in a fanciful short story, but later overtly incorporated into *The War of the Worlds* and covertly into *The Time Machine* (1895) and *The First Men in the Moon* (1901).[25] The man of the year million consisted of little else but a hand and an enormous brain; all other physical organs had atrophied, or they were no longer needed. Wells's Martians and his Grand Lunar are realizations of this idea of the future man; his degenerate Eloi and Morlocks, who flourish a little short of a million years hence, are the products of an equally drastic course of physical evolution. After 1901, Wells abandoned these fictions with their allegorical glimpses of the far future and, as Van Wyck Brooks put it, "domesticated himself in his own planet and point of time"; but he still based his fiction on the use of models or specimens. Both Kipps and Mr Polly are occasionally seen in this light. In *Tono-Bungay* the social analysis is held together by Wells's modelling vision of Bladesover, the country-house on the Kentish Downs, as a "complete authentic microcosm" of traditional English society.[26] George Ponderevo's understanding of his world is decisively influenced by his upbringing in the servants' quarters of the great house. His later escapades, which include a buccaneering mission to tropical Africa, only confirm his hunch that England and its Empire are permeated by the "Bladesover system," which is now subject to spreading hypertrophy and decay. Scientific invention, represented by George's experimental naval craft and his flying machines, and commercial enterprise, represented by his uncle's fraudulent ventures, flourish as best they can in the interstices of this structure. The voyage down the Thames at the end of the novel is a symbolic rejection of an England weighed down by its history.

Ponderevo is not a limited hero like Kipps and Mr Polly, but a first-person narrator whose confused strivings after the ideal of disciplined imagination evidently reflect Wells's own. He finally professes his faith in science, "the remotest of mistresses," but in another respect he is a shameless adventurer, as Wells himself was.[27] In a 1911 article, Wells described the literary life as "one of the modern forms of adventure. Success with a book . . . means in the English-speaking world not merely a moderate financial independence but the utmost freedom of movement and inter-course. One is lifted out of one's narrow circumstances into familiar and unrestrained intercourse with a great variety of people. One sees the world."[28] After *Tono-Bungay*, Well's fictional heroes also tend to become promiscuous globe-trotters. In part, this is an expression of the needs of disciplined imagination: the prospect of life unfolded in the picture of Education in *Mr Polly* must be tested and known at firsthand by a hero aspiring to full consciousness, as George Ponderevo does. But it also reflects the change in Wells's own experience that accompanied his growing pros-perity, and the principle of opportunism inherent in autobiographical fic-tion. Five years after *Tono–Bungay* with its concern with the Condition of England, Wells collected his essays on a variety of subjects and gave them the appropriate title *An Englishman Looks at the World*.[29]

Certainly he never wrote a conventional travel book. The things that he saw in foreign countries were, on the whole, like the boat riding at anchor off Sandgate beach; they had little appeal to the irresponsible side of his imagination. Where he was most effective was in pioneering a certain kind of twentieth-century reportage, in which travel takes the shape of a frustrated but ever-hopeful pilgrimage toward an imagined political new world. Both *The Future in America* (1906) and *Russia in the Shadows* (1920) are books of this sort, as are the much less durable *Washington and the Hope of Peace* (1922), *Stalin-Wells Talk* (1934), *The New America: The New World* (1935), and his record of a visit to Australia, *Travels of a Republican Radical in Search of Hot Water* (1939). All these books recount the official travels of one who was already a public figure. Given the moral conventions of the time it is useless to speculate on what sort of book *The Future in America* might have been had Wells felt able to confess (as he did in his posthumous volume of autobiography) that, after calling on Theodore Roosevelt at the White House, he had spent the rest of a highly satisfying afternoon with a black prostitute. Similarly, his writings about the Soviet Union were never complicated by any analysis of his close relationship with Moura Budberg, who is now alleged to have been a Soviet agent.[30]

One clear advantage of *The Future in America* and *Russia in the Shadows* over the novels that were contemporary with them is their sense of history in the making. Nothing that Wells could put into a novel could rival his account of a meeting with Lenin, "The Dreamer in the Kremlin," even though *The World Set Free* (1914) had already included a portrait of a fictional world leader. Nevertheless, the fictional pilgrimage or Grand Tour is an integral part of the sequence of works, beginning with *Ann Veronica* (1909) and *The New Machiavelli* (1911), which are usually known as the "discussion novels" (or "prig novels"), but which increasingly become globe-trotting novels as well. In *Ann Veronica* the love between the heroine and Capes is sealed by a wordy honeymoon in the Swiss Alps. *The New Machiavelli* returns to the Alps for a high-minded walking tour. *Marriage* also features an Alpine walking tour, which proves inconclusive since the hero and heroine, Trafford and Marjorie, abandon their knapsacks in order to enjoy the hospitality of a rich industrialist with a Swiss holiday villa. The Traffords finally decide to make a further pilgrimage to a real wilderness, spending a winter in a hut in the midst of Labrador. Like the Samurai (the ruling class of *A Modern Utopia*) the Traffords need to survive the test of the wilderness in order to emerge as "new selves" capable of fulfilling their true human potential. International tourism takes the place of the call of the wild in *The Passionate Friends* (1913) and *The Research Magnificent* (1915), turgid books that would hardly claim our attention were it not that they could be described as the ultimate globe-trotting novels. Stratton, in *The Passionate Friends*, goes out to volunteer in the South African War. "It isn't my business to write here any consecutive story of my war experiences," he tells the reader, and it is his general reflections on Imperialism and world development which take over the narrative.[31] Later, in similarly reflective vein, he visits the United States and India (where he survives an encounter with a tiger), and becomes a frequent traveler to world peace conferences. Much the same is true of Benham in *The Research Magnificent*. After the usual forays to Switzerland and Italy, he decides to go around the world, and sets out for Moscow. Russia to him is merely Britain writ large:

> St Petersburg upon its Neva was like a savage untamed London on a larger Thames; they were sea-gull-haunted tidal cities, like no other capitals in Europe. . . . Like London it looked over the heads of its own people to a limitless polyglot empire. . . . One could draw a score of such contrasted parallels. And now [Russia] was in a state of intolerable stress, that laid bare the elemental facts of a great social organisation. It was having its South African war, its war at the other end of the earth, with a certain

228

defeat instead of a dubious victory.[32]

Once in Moscow, Benham is involved in "trying to piece together a process, if it was one and the same process, which involved riots in Lodz, fighting at Libau, wild disorder at Odessa, remote colossal battlings in Manchuria, the obscure movements of a disastrous fleet lost somewhere now in the Indian seas."[33] Meanwhile, his companion is enjoying a love affair with a half-Russian, half-English mistress picked up at the Cosmopolis Bazaar. Later, in an increasingly fragmentary narrative, Benham's curiosity takes him to India and China and then to South Africa, where he is killed in a riot.

In these unsatisfactory works Wells was pioneering an idea of "revolutionary sightseeing"—a kind of compulsive traveling to the world's trouble spots—which was later to become a regular feature of twentieth-century life and a source of livelihood for writers and journalists. In *Mr Britling Sees It Through* (1916), however, the Great War virtually confines the Wellsian hero to his country home in Essex, and the novel is all the better for it. For Wells himself, the best that can be said about his sequence of voraciously philosophical globe-trotting protagonists is that they turned his mind to the writing of history. *The Outline of History* (1920) was a textbook for the world, intended to supplant the popular nationalistic versions of history which, Wells believed, had contributed to the catastrophe of the First World War. *The Outline* and its successor *A Short History of the World* (1922) have been frequently revised and remain in print today. They were deservedly successful and continue to be valued—by Asian and American readers, among others—for their attempts to displace the Eurocentric versions of world history engendered by the age of Imperialism. Wells's study of history was part and parcel of his search for a new world, since *The Outline* ends not in the present but in the near future. In its original serial publication, the final volume had on its cover a map of the world without political subdivisions, entitled "The United States of the World." The text of the first edition concluded with a discussion of "The Next Stage in History," arguing for the necessity of a federal world government.[34]

If historiography eventually becomes prophecy in Wells's hands, travelogue turns much more quickly into utopian vision. In *A Modern Utopia* (1905), once again, he starts out by using the device of the Swiss walking tour. As the narrator and his disputatious companion descend the pass leading from Switzerland into Italy, they find that they have been miraculously transported into Utopia, which is represented as a parallel planet at the other end of the galaxy. In due course their explorations bring them back to utopian London, and, standing in the dignified colonnade which in Utopia corresponds to Trafalgar Square, they find themselves back with a bump in the familiar city. This is not quite the end of the story, for Wells's narrator,

229

traveling away on the top of a bus, imagines the figure of an apocalyptic angel towering over the Haymarket. The trumpet sounds, and he has a momentary vision of "a world's awakening" to the utopian spirit.[35] This vision from the Book of Revelations helps to explain why Wells's literal travel writing is so much feebler than his accounts of journeys in time or to parallel worlds. What interested him was the cosmic promise of a new world, which only the imagination could envisage.

Though *The Future in America* (1906) describes a visit to an actual New World and contains some memorable impressions of the United States, it begins and ends with those open-ended invocations of the future—the speculative metaphors, the sense of continuing inquiry, the sentences tailing off into suspension-points—which by 1906 were becoming Wells's trademark. The first chapter, an essay in self-reflection called "The Prophetic Habit of Mind," poses the aim of his transatlantic voyage as being "to find whatever consciousness or vague consciousness of a common purpose there may be, what is their Vision, their American Utopia, how much will there is shaping to attain it."[36] At the end of the book, after an exhilarating but inconclusive search for that consciousness of purpose, he describes another apocalyptic fantasy, though this time it is subdued and understated. Looking back at the skyscrapers which by then composed the New York skyline, he is irresistibly reminded of "piled-up packing cases outside a warehouse." Out of them presently will come "palaces and noble places," and "light and fine living," or so he affirms.[37] Though his rhetoric is commonplace, the packing-case metaphor momentarily shows Wells at his best, in that it deconstructs New York's monumental buildings and treats them as mere disposable containers for the energies of the people who live in them. Although the New World (inevitably) failed to measure up to Wellsian standards, it remained a source of possible new worlds.

The ultimate New World was the conquest of space. However, space travel tends to appear only as a source of rhetorical uplift at the end of his works—as in *The Discovery of the Future* (1902), for example, or in *Things to Come*. The popular fiction of space adventure was, on the whole, the creation of writers such as Edgar Rice Burroughs, together with a generation of Wellsian followers such as Robert A. Heinlein and Arthur C. Clarke (not forgetting his more intellectual disciple Olaf Stapledon). *The First Men in the Moon* (1901) was Wells's one contribution to this genre. Here the lunar landscapes, in which T.S. Eliot found "imagination of a very high order,"[38] are dominated by the shock of the sunrise after the long lunar night, and by the hectic growth of vegetation in the low lunar gravity. This is, quite explicitly, a description of a strange new world, compared by the narrator to the miracle of the Creation. The two lunar travelers, Bedford and Cavor—

the prospector and the disinterested explorer—are engaged on a conventional Imperial mission. At times the obstacles they face are comparable to those of a desert or tropical jungle, but the forms and colors of the vegetation that composes the lunar landscape are also reminiscent of an enormously distended suburban rock garden. It is also a pastoral world, with Selenite shepherds tending the flocks of grazing mooncalves. All this would suggest that Bedford and Cavor have journeyed from Kent, the "Garden of England," to a place which, however strange, is another garden world. More familiar garden-worlds are found in some of Wells's visions of the future, from *The Time Machine* to his later utopian books. In 1924 he confessed that his imagination took "refuge from the slums of to-day in a world like a great garden, various, orderly, lovingly cared-for."[39] This is plainly an Edenic vision, but it is also a very English one, belonging to the world of (for example) William Morris's *News from Nowhere* (1891) and Ebenezer Howard's *Garden Cities of To-Morrow* (1902).

In *The Time Machine*, and again in *Men Like Gods* (1923), the urban and industrial landscape has reverted to that of a country park, such as Uppark in Sussex where Wells's mother was housekeeper and where his father had worked as a gardener. The landscape of *The Time Machine* can still be appreciated by a visitor to Richmond Park in Surrey. In the 800,000 years that have elapsed between the nineteenth century and the period of the Eloi and Morlocks, a garden-city civilization has come and gone, leaving behind it the great ruined buildings in which the decadent Eloi huddle for shelter. *Men Like Gods* presents another parkland scene, complete with distant snow-capped mountains and tame wildcats. A party of Earthlings accidentally enters this paradise as a result of a utopian experiment in rotating time-space planes; and they do so from a location in the Home Counties, as they are motoring away from London on the Great West Road. Here, and elsewhere, Wells's anxiety to put southern England behind him is only equalled by his determination that the "garden world" that replaces it will be reminiscent of the England his characters have left. At such time he reminds us of an earlier English visionary writer, William Blake, who sang of building Jerusalem, the perfect city, in the English countryside, and who saw "Another England" in "The Crystal Cabinet":

> Another England there I saw
> Another London with its Tower
> Another Thames and Other Hills
> And another pleasant Surrey Bower (ll. 9–12)

However, there is another analogy that fits Wells's vision of new worlds, and that is with the pioneer's or colonist's mentality. If the colonist's first priority is to destroy all links with the homeland, the second priority is to construct a new settlement that at once fulfills the promise of a better society and serves as a memorial to the homeland. Some such logic must have inspired the New England settlers, as well as all other pioneers in the white-settler lands who gave to their new homes in the wilderness familiar British and European names. Wells's deliberate cosmopolitanism and his proclamation of world citizenship remain important and worthwhile ideals, but his deeper affinity is with the New World spirit, even though he himself was never tempted to emigrate. The destruction that many emigrants must have wished on the homelands they were leaving is enacted in *The Time Machine*, *The Invisible Man*, and *The War of the Worlds* as well as in *Mr Polly*, where the hero's failed suicide attempt fortuitously succeeds in burning down Fishbourne High Street. In *Tono-Bungay*, the Wellsian hero George Ponderevo is a spiritual emigrant who passes the whole of English society in review before bidding it an embittered farewell. But Wells's novels also anticipate the dreamed-of return to and utopian reconstruction of the homeland, and it is this compound sense of otherness superimposed on Englishness, of an old world irresistibly giving way to a new one, which inspires his most imaginative writing.

NOTES

The works of Wells are quoted from the following editions: *Experiment in Autobiography* (London: 1966); *The Future in America* (London: 1906); *Tono-Bungay* (London: 1909); *Kipps* (London: 1905); *The Outline of History* (London: 1920); *The History of Mr Polly* (London: 1910); *A Modern Utopia* (London: 1905); *Complete Short Stories* (London: 1927); *The Country of the Blind and Other Stories* (London: n.d. [1911]); *The Passionate Friends* (London: 1913); *The Research Magnificent*, Essex edition (London: 1927); *A Year of Prophesying* (New York: 1925).

1. H.G. Wells, *H.G. Wells in Love*, ed. G.P. Wells (London: 1984), 235.
2. See David C. Smith, *H.G. Wells: Desperately Mortal* (New Haven, CT, and London: 1986), 333.
3. Anthony West, *H.G. Wells: Aspects of a Life* (London: 1984), 132.
4. See, e.g., Leon Stover, *The Prophetic Soul: A Reading of H.G. Wells's Things to Come* (Jefferson, NC, and London: 1987), passim—though Stover also claims major intellectual and artistic significance for *Things to Come*.
5. See Robert Colls and Phillip Dodd, eds., *Englishness: Politics and Culture 1880–1920* (London: 1986).
6. On Comte, see H.G. Wells, "The So-Called Science of Sociology," in *An Englishman Looks at the World* (London: 1914), 192–93. On Marx, see Wells, *Experiment in Autobiography*, 1.263–4.
7. *Autobiography*, 1.54.
8. *The Future in America*, 24–25.

9. *Tono-Bungay*, 3.
10. Ibid., 492, 493.
11. Van Wyck Brooks, *The World of H.G. Wells* (London: 1915), 178.
12. *Kipps*, 5.
13. *The Outline of History*, 608.
14. Cf. Richard Brown, "Little England: On Triviality in the Naive Comic Fictions of H.G. Wells," in *Cahiers Victoriens et Edouardiens* (October 30, 1989), pp. 55–66.
15. *The History of Mr Polly*, 16–17.
16. *The Outline of History*, 529, and *Autobiography*, 1.80.
17. *A Modern Utopia*, 72.
18. Brooks, *The World of H.G. Wells*, 134.
19. *Complete Short Stories*, 307.
20. *The Country of the Blind and Other Stories*, iv.
21. *Autobiography*, 2.543.
22. Michael Craper, "The Martians in Ecuador," *Wellsian*, 5 (Summer 1982), 35–36.
23. "Scepticism of the Instrument," in *A Modern Utopia* (November 16, 1893), 376.
24. *Autobiography*, 2.619.
25. "The Man of the Year Million," *Pall Mall Budget* (November 16, 1893), 1796–97.
26. *Tono-Bungay*, 9.
27. Ibid., 346.
28. "Mr Wells Explains Himself," quoted in Patrick Parrinder, *H.G. Wells* (Edinburgh: 1970), 5–6.
29. See David Lodge, "*Tono-Bungay* and the Condition of England," in *Language of Fiction* (London: 1966), 214–42.
30. See Anthony West, *H.G. Wells: Aspects of a Life*, 143–46.
31. *The Passionate Friends*, 105.
32. *The Research Magnificent*, 258–59. The final suspension points are Wells's own.
33. Ibid., 262.
34. *The Outline of History*, 601–8.
35. *A Modern Utopia*, 369.
36. *The Future in America*, 21.
37. Ibid., 358, 359.
38. T.S. Eliot, "Wells as Journalist," in *H.G. Wells: The Critical Heritage*, ed. Patrick Parrinder (London and Boston: 1972), 320.
39. *A Year of Prophesying*, 351.

SELECT BIBLIOGRAPHY OF SECONDARY WORKS

1. Travel and Travel Writing

Bethke, Frederick John. *Three Victorian Travel Writers: An Annotated Bibliography of Criticism on Mrs. Frances Milton Trollope, Samuel Butler, and Robert Louis Stevenson*. Boston: 1977.

Cooke, Stenson. *This Motoring, Being the Romantic Story of the Automobile Association*. London: 1931.

Kerr, Barbara. *The Dispossessed: An Aspect of Victorian Social History*. New York: 1974.

Pimlott, J.A.R. *The Englishman's Holiday: A Social History*. London: 1947.

Pudney, John. *The Thomas Cook Story*. London: 1953.

Stevenson, Catherine. *Victorian Women Travel Writers in Africa*. Boston: 1982.

Swinglehurst, Edmund. *The Romantic Journey: The Story of Thomas Cook and Victorian Travel*. London: 1974.

Tinling, Marion. *Women into the Unknown*. New York: 1989.

2. Imperialism

Brantlinger, Patrick. *Rule of Darkness: British Literature and Imperialism 1830–1914*. Ithaca: 1988.

Darby, Philip, *Three Faces of Imperialism*. New Haven, CT, and London: 1987.

Dilke, Sir Charles. *The British Empire*. London: 1899.

Goonetilleke, D.C.R.A. *Developing Countries in British Fiction*. London and Basingstoke, Eng.: 1977.

Green, Martin. *Dreams of Adventure, Deeds of Empire*. London: 1980.

Hall, Winifred. *The Overseas Empire in Fiction*. London: 1942.

Hobson, J.A. *Imperialism: A Study*. London: 1905.

———. *The Psychology of Jingoism*. London: 1901.

Howe, Susan. *Novels of Empire*. New York: 1949.

Kiernan, V.G. *The Lords of Human Kind: European Attitudes Towards the Outside World in the Imperial Age*. London: 1969.

Koebner, Richard, and Helmut Dan Schmidt. *Imperialism: The Story and Significance of a Political Word*, 1840–1960. Cambridge, Eng.: 1965.

Raskin, Jonah. *The Mythology of Imperialism*. New York: 1971.

Sandison, Alan. *The Wheel of Empire: A Study of the Imperial Idea in Some Late Nineteenth and Early Twentieth Century Fiction*. New York: 1967.

Seely, J.R. *The Expansion of England*. London: 1883.

Thornton, A.P. *The Imperial Idea and Its Enemies*. London: 1959.

3. Imperial Architecture

Davies, Philip. *Splendours of the Raj*. London: 1985.
Freeland, J.M. *Architecture in Australia: A History*. Melbourne: 1968.
Metcalf, Thomas R. *An Imperial Vision: Indian Architecture and Britain's Raj*. Berkeley and Los Angeles: 1989.
Picton-Seymour, Desirée. *Victorian Buildings in South Africa 1850–1910*. Cape Town: 1977.

4. France.

Campos, Christophe. *The View from France*. London and New York: 1965.
Crossley, Ceri, and Ian Small, eds. *Studies in Anglo-French Relations*. London and Basingstoke, Eng.: 1988.
Howarth, Patrick. *When the Riviera Was Ours*. London and Henley, Eng.: 1977.
Starkie, Enid. *From Gautier to Eliot: The Influence of France on English Literature 1851–1939*. London: 1960.

5. Italy

Churchill, Kenneth. *Italy and English Literature 1764–1930*. Totowa, NJ: 1980.
Martini, Paola Maria. "The Image of Italy and the Italian in English Travel Books: 1800–1901." Diss. U. of Washington 1983.

6. The Mediterranean

Pemble, John. *The Mediterranean Passion*. Oxford: 1987.

7. Switzerland

De Beer, Gavin Rylands. *Travellers in Switzerland*. London and New York: 1949.
Lunn, Arnold. *Switzerland and the English*. London: 1944.
Nicolson, Marjorie Hope. *Mountain Gloom and Mountain Glory: The Development of the Aesthetics of the Infinite*. Ithaca, NY, and New York: 1959.

8. Germany

Argyle, Gisela. *German Elements in the Fiction of George Eliot, Gissing and Meredith*. Frankfurt, Bern, and Cirencester: 1979.
Firchow, Peter. *The Death of the German Cousin*. Lewisburg, PA: 1986.
Grey, Ronald. *The German Tradition in Literature*. Cambridge, Eng.: 1965.
Mander, John. *Our German Cousins*. London: 1974.

9. Scandinavia

Herford C.H. *Norse Myth in English Poetry*. [Folcroft] 1970.
Wright, H.G. *Studies in Anglo-Scandinavian Literary Relations*. Bangor, Wales: 1919.

10. Russia

Brewster, Dorothy. *East-West Passage: A Study in Literary Relationships*. London: 1954.
Cross, Anthony G. *The Russian Theme in English Literature from the Sixteenth Century to 1980*. Oxford: 1985.

11. The Middle East and Egypt

Daniel, Norman. *Islam and the West: The Making of an Image*. Edinburgh: 1960.
Mansfield, Peter. *The British in Egypt*. London: 1971.
Said, Edward. *Orientalism*. London: 1978.

12. Africa

Grey, Stephen. *Southern African Literature*. Totowa, NJ: 1979.
Hammond, Dorothy, and Alta Jablow. *The Myth of Africa*. New York: 1977.
Killam, G.D. *Africa in English Fiction 1874–1939*. Ibadan: 1968.
Malherbe, V.C. *Eminent Victorians in South Africa*. Cape Town, Wynburg, and Johannesburg: 1972.
Robinson, Ronald, and J. Gallagher, with Alice Denny. *Africa and the Victorians: The Climax of Imperialism in the Dark Continent*. New York: 1961.
van Wyk Smith, Malvern. *Drummer Hodge: The Poetry of the Anglo-Boer War (1899–1902)*. Oxford: 1978.

13. India

Edwardes, Michael. *Bound to Exile: The Victorians in India*. London: 1969.
Greenberger, Allen. *The British Image of India: A Study in the Literature of Imperialism 1880–1960*. London: 1969.
Kincaid, Dennis. *British Social Life in India, 1607–1937*. London: 1939.
Singh, Bhupal. *A Survey of Anglo-Indian Fiction*. London: 1934.

14. The Far East

Edwardes, Michael. *East-West Passage: The Travel of Ideas, Arts and Inventions between Asia and the Western World*. London: 1971.
Miner, Earl. *The Japanese Tradition in British and American Literature*. Princeton: 1958.

15. The Pacific and Australia

Friedrich, Werner. *Australia in Western Imaginative Prose Writings 1600–1960*. Chapel Hill, NC: 1967.

Lansbury, Coral. *Arcady in Australia: The Evocation of Australia in Nineteenth-Century English Literature*. Melbourne: 1970.

Pearson, Bill. *Rifled Sanctuaries: Some Views of the Pacific Islands in Western Literature to 1900*. Auckland, Oxford, and New York: 1984.

16. America

Allen, Walter, ed. *Transatlantic Crossing: American Visitors to Britain and British Visitors to America in the Nineteenth Century*. London: 1971.

Conrad, Peter. *Imagining America*. New York: 1980.

Gordon, George Stuart. *Anglo-American Literary Relations*. London and New York: 1942.

Rapson, Richard. *Britons View America: Travel Commentary, 1860–1935*. Seattle and London: 1971.

17. War

Buitenhuis, Peter. "Writers at War: Propaganda and Fiction in the Great War." *University of Texas Quarterly* 45: 277–94.

Eby, Cecil. *The Road to Armageddon*. Durham, Eng., and London: 1987.

Fussell, Paul. *The Great War and Modern Memory*. New York: 1977.

Melchiori, Barbara Arnett. *Terrorism in the Late Victorian Novel*. London, Sydney, Dover, NH: 1985.

Rutherford, Andrew. *The Literature of War*. London and Basingstoke, Eng.: 1978.

A GUIDE TO AUTHORS

BARING, Maurice (1874–1945), made many journeys to Russia between 1904 and 1916, first as a journalist, then as a Russophile (*With the Russians in Manchuria*, 1905; *A Year in Russia*, 1907; *Russian Essays and Stories, 1908; The Russian People*, 1911).

BELL, Gertrude (1868–1926), 1888–89 to Bucharest; 1892–93 to Teheran (*Safar Nameh: Persian Pictures*, 1894); 1897 to Berlin; 1897–99 voyage around the world; 1899 winter in Jerusalem; 1901, 1902, and 1904 trips to the Alps; 1905 departs from Jerusalem for Syria and Cilicia and to Konia in Asia Minor (*The Desert and the Sown*, 1907); 1907 explores the Hittite and Byzantine archeological site of Bin-bir-kilisse, near Isaura; 1909 journeys down the Euphrates from Aleppo to Ukhaidir, an early Islamic palace near Kerbela, returning through Baghdad and Mosul to Asia Minor; 1911 returns to explore the Ukhaidir; 1913 attempts a journey into central Arabia, but makes it only as far as Hail; 1914 service with the Red Cross in Boulogne; 1915 travels to Cairo to help form an Arab intelligence bureau; 1916 travels to Delhi and Basra on government business; 1917 moves to Baghdad, serving as Oriental secretary to Sir Percy Cox, civil commissioner, and to his successor, Sir Arnold Wilson; 1926 dies in Baghdad.

BELLOC, Hilaire (1870–1953), early childhood spent partly in La Celle St. Cloud (France) and partly in Wimpole Street (London); 1877 settles permanently in England; 1887 enters the Naval Class of the College Stanislas, in Paris, leaving after two months; 1888 extended tour of Ireland; 1890–91 to Brittany to cover the French general election for *Paternoster Review*, then to the Auvergne; 1890 travels to America in pursuit of Elodie Hogan (Manhattan, Philadelphia, Cincinnati, then to San Francisco, usually penniless and often on foot); 1891 joins the French army–10th Battery of the 8th Regiment of Artillery at Toul; 1893 holiday in Scandinavia with Basil Blackwood; 1896 second trip to America, arriving in New York and traveling to San Francisco by train; he marries Elodie Hogan in Napa; 1896 holiday with Elodie in Normandy; 1897 lecture tour of America; 1897–98 again to America; 1898 to Paris; 1900 again to Paris to research for a history of the city (*Paris*, 1900); 1900 holiday at Portofino, Italy, and wanderings about northern Italy; 1901 travels by foot from Toul, along the Moselle, over the Alps and down through Tuscany to Rome (*The Path to Rome*, 1902); 1905 travels in North Africa; frequent travels to France, including a bicycle holiday with Elodie in 1904 (*Esto Perpetua*, 1906); 1910 visits Ireland; 1912 short tour of France, Belgium, and Germany; 1912 trip to Moscow to research for *Pall Mall Gazette* articles on Napoleonic battlefields; 1913 trip to France with George Wyndham; 1914 travels abroad after Elodie's death, including Rome (where he had an audience with the pope), Naples, Sicily, Tunisia, Marseilles, Provence, and Lyons; over twenty trips abroad in 1915–16; 1915 to France (Bologne, Ypres, and Cassel); 1916 visits Italian front and has a papal audience arranged by the British Foreign Office in the Vatican; 1923 visits America for a lecture tour (New York, Cincinnati, Chicago, and Des Moines); 1923 to Barcelona, Spain; 1924 to Rome to write about Mussolini and the "new Italy"; 1927 at the edge of the Sahara Desert in Africa, writing *James II*; 1933 visits Italy, where he meets Evelyn Waugh in Portofino near Genoa; 1935 lecture tour of the Eastern Seaboard of the United States; return trip includes Cuba, Spain, and the Holy Land; 1936 Continental tour, including Berlin, Poland, Vienna, Burgundy, and Paris; returns to Paris in April; 1937 U.S. lecture tour; 1938 six months in Paris; later that year a Scandinavian tour; 1939 visits Rome to cover funeral of Pope Pius XI for Hearst Press, followed by six weeks in Paris; later that year he visits France and sees his old regiment in the Maginot Line with Henri Matisse; also to Belgium; 1940 spring visit to Paris to look for his daughter Elizabeth.

BENNET, Enoch Arnold (1867–1931), 1903 moves to France, where he lives until 1912 (*The Old Wives' Tale*, 1908); 1908 three months in Switzerland; 1911 six weeks in America (*Those United States*, 1912); 1914 voyage along the North Sea coast of Europe (*From the Log of the Velsa*, 1914); 1915 tours Western Front for the government; 1918 becomes director of propaganda, Ministry of Information for the War Office in France; 1930 travels to France, where he contracts typhoid fever.

BISHOP, Isabella (née Bird) (1831–1904), 1874 to Australia, New Zealand, the Sandwich Islands (Hawaii) (*Six Months in the Sandwich Islands*, 1875) and the United States (*A Lady's Life in the Rocky Mountains*, 1879); 1878 to Japan via America (*Unbeaten Tracks in Japan*, 1880); 1879 from Japan to Hong Kong, Canton, and the Malay States (*The Golden Chersonese*, 1883); 1884–85 on the Riviera and in Switzerland with her ailing husband; 1887 to Ireland; 1889–90 to Kashmir, Tibet (*Among the Tibetans*, 1894), Persia, Kurdistan (*Journeys in Persia and Kurdistan*, 1891), and Turkey; 1894–97 to Korea (*Korea and her Neighbours*, 1897), China (*The Yangtze Valley and Beyond*, 1899), Russia, and Japan via Canada; 1901 to Morocco.

BLUNT, Wilfrid Scawen (1840–1922), 1858 enters the diplomatic service and for the next eleven years serves as secretary at various posts, including Athens, Paris, and Madrid; 1863 to Bordeaux; 1876–77 travels with his wife to the Nile Delta and the Sinai; 1877–78, explores Mesopotamia along the Tigris and Euphrates; 1878–79 trip to central Arabia to the area of Nejd (probably the first European Christians to enter the area without disguise), then north toward Persia; 1879 arrives in Baghdad then Bushire on the Persian Gulf; a month later in India; 1882 purchases Sheykh Obeyd, the Blunts's Egyptian home, near Heliopolis (*The Future of Islam*, 1882; *The Wind and the Whirlwind*, 1883); 1883 trip to India, touring the cities (*Ideas about India*, 1885); 1886 two trips to Ireland (*The Land War in Ireland*, 1912) and one to Rome, where he had an audience with Pope Leo XIII; 1887 to Egypt for five months; later that year to Ireland; 1897 travels from Cairo to the Oasis of Siwa in the Western Desert disguised as an Arab.

BROOKE, Rupert (1887–1915), 1905 and 1906 travels to Italy; 1908(?) travels to Germany; 1912 travels for the *Westminster Gazette* across the United States and Canada, leaving from San Francisco and touring the South Sea Islands (Hawaii, Samoa, Fiji, and Tahiti) (*The South Seas* in *Poems 1911–14*, 1918); 1914 joined the Royal Navy Volunteer Reserve and served in Belgium; 1915 dies in Cairo.

BUCHAN, John (1875–1940), 1901–1903 in South Africa on the staff of High Commissioner Lord Alfred Milner; 1902 to western frontiers of the Transvaal; 1903 to the Wood Bush northeast of Johannesburg in January, to Swaziland in March, and to the Rhodesian border in July (*The African Colony*, 1903; *Prester John*, 1910); 1904 tours the Swiss Alps; 1905 and 1906 tours the French Alps; 1907 honeymoon with Susan Grosvenor to Italy (Achensee, Cortina, and Venice); 1910 with his wife, travels on the Orient Express to Constantinople, including a cruise on the Aegean and a visit to Athens (*The Dancing Floor*, 1926); 1911 holiday in Bavaria; 1912 fishing holiday in Norway; 1913 cruise to the Azores; 1910 to Dublin to visit wife's uncle, Sir Neville Lyttelton; 1915 travels to France for the War Office, making visits to the front, 1916 working in the French Foreign Office as a major in the intelligence corps; 1924 visits Canada and the United States; 1935 created Baron Tweedsmuir of Elsfield and governor-general of Canada; 1937 trip to the Arctic (*Sick Heart River*, 1941).

CHILDERS, Erskine (1870–1922), young adulthood sails the Channel and the North Sea during holidays, including German, Danish, and Baltic coasts; 1899–1900 serves in the Boer War in the City Imperial Volunteer battery of the Honourable Artillery Company; 1903 travels to Boston with the Honourable Artillery Company; 1904 returns to London; from 1904 yachting chiefly in the Baltic on *Asgard*; during World War I, reconnaissance work on the seaplane carrier HMS *Engandine* and other posts in the Royal Naval Air Service; 1919 visits Paris with the Irish republican envoys to the Versailles Conference;

same year, he settles with his family in Dublin.

CONRAD, Joseph (1857–1924), born Józef Teodor Konrad Korzeniowski in Russian Empire to Polish nationalists; 1862 moves with his parents to exile in northern Russia before moving to the Polish area of Austria-Hungary; 1874–78 living in Marseilles, working on sailing ships—during these years he made voyages to the West Indies (*Romance* [with Ford], 1903; *Nostromo*, 1904); 1879 moves to England as an ordinary seaman on a British steamer, making voyages primarily to the Far East, working his way up to captain (*Almayer's Folly*, 1895; *An Outcast of the Islands*, 1896; *Lord Jim*, 1900; *Typhoon*, 1902); 1890 on a Congo steamer ("An Outpost of Progress," 1898; *Heart of Darkness*, 1902); 1896 honeymoon in Brittany.

CUNNINGHAME GRAHAM, Robert Bontine (1852–1936), 1868 at school in Brussels; 1870–72 working in Argentina; 1873–74 working in Paraguay; 1876–77 working in Uruguay, Brazil, and Argentina; 1879–81 in New Orleans, Texas, and Mexico (numerous stories come from his experiences in Latin America, as well as seven volumes of the history of the Spanish conquest of the area); 1894 prospects for gold in Spain; 1897 travels in Morocco, disguised as an Arab (*Mogreb-el-Acksa*, 1898); 1914 to Argentina to buy horses for the army; 1916–17 surveying cattle resources of Colombia.

DOUGLAS, Norman (1868–1952), 1883–89 attends Gymnasium at Karlsruhe, Germany; 1888 first visit to Capri; 1889 visits Paris and takes a bicycle tour in the French countryside; 1889 trips to Iceland and the Hebrides; 1891 visits Lipari Islands; 1892 visits Greece, Santorini, Malta, Filfa, Brindisi, returning to Athens via Corfu and Patras; also visits Naples; 1893 visits Italy and Greece; 1894–96 in St. Petersburg, Russia, as an attache in the British Embassy; 1894 spends his leave touring Finland; 1895 spends his leave traveling in Asia Minor; in June, back in the Lipari Islands, then Genoa, Venice, and finally the Vorarlberg; 1896 leaves the Foreign Service and goes to Italy; 1897 settles on the Bay of Naples in the Villa Maya, four miles from the city; 1897 (December) sets out for first visit to India; 1899 visits Tunisia; 1900 visits India again and Ceylon; 1901 visits England and Scotland; 1902 to Bludenz, Austria; 1903 moves to Capri, where he is based for the next ten years, interspersed with trips to London; 1907 first of many visits to Calabria (*Old Calabria*, 1915); 1909 visits Tunisia to collect material for *Fountains in the Sand* (1912); 1914 visits Capri; 1916 visits Italy (alone); in Capri May–August working on *South Wind* (1917); 1917 leaves England to settle in Florence in February; wanders through the Latium beginning in May; two weeks in Rome in August, then to Paris in October, where he lived for a year; 1918 visits St. Malo, leaves Paris for Menton in December; 1919 arrives in Florence in September, which was to be his headquarters for eighteen years; 1920 trip to Greece, 1921 and 1922 trips to the Vorarlberg; 1937 leaves Florence and moves to the South of France, where he lives at Vence and Antibes until 1940; 1940 in Lisbon; 1941 in London; 1946 back at Capri where he lives until his death.

FLECKER, James Elroy (1884–1915), 1910 enters consular service, posted to Constantinople; 1911 transferred to Smyrna, then to Beirut (*The Golden Journey to Samarkand*, 1913); health breaks down, in Swiss sanitarium (*Hassan*, 1922).

FORD, Ford Madox (1873–1939), born Ford Madox Hueffer; 1889 trip to Paris to visit Hueffer relatives; 1899 holiday to Belgium with the Conrads; 1904 trip to Germany for health reasons, then to Basel, Switzerland, to write a life of Holbein (1905); 1906 lecture tour of America (New York, Philadelphia, Newport, Boston); 1910 to Germany with Violet Hunt, settling for a time at Giessen (*High Germany*, 1912; *The Good Soldier*, 1915); 1911 meets Violet Hunt at Rheims and they travel together to Belgium then to Rome in late autumn; 1916 to France as an officer in the British Army; 1922 moves to France with Stella Bowen, first to Paris then south to Cap Ferrat on the Riviera; 1923 to Provence, from which he visited Arles, Pont du Gard, and Nimes, moving to St. Agreve for the summer, then back in Paris in September where he became editor of the *Transatlantic Review*; winters of 1924–25 and 1925–26 living at Toulon; 1926 and 1927 to America for lecture tours of the east and the midwest; 1928 to America to arrange for publishing contracts; later that year to Monte Carlo and Corsica to gather material for *A Little Less than Gods*; 1929 trip to New York, returning to Paris after a brief stop in London; 1930 lecture tour in U.S.; summer of that year in south of France, outside Toulon; 1930–32 based in Paris, summering at Cap Brun; 1932 visits Ezra Pound at Rapallo, Italy; 1934 trip to London for publication of *It Was the Nightingale*; returns to Toulon, then Christmas in Paris; 1934–35 to America, first to New York on business, then to Tennessee in the spring to visit Allen Tate and Caroline Gordon and to Louisiana State University for the Southern Writers' Conference; 1935 returns to Cap Brun; 1935 to Switzerland; 1936 visits London; 1936–37 to New York, then to Tennessee to visit Tate-Gordons, where he began *March of Literature*; 1937 first at Olivet College in Michigan for a writers' conference, then to University of Colorado, back to Olivet in the autumn and returning to Paris in the winter; 1938 spring at Olivet, lecture tour in the summer, and October in New York; 1939 begins return to France, but becomes ill on the journey and dies of heart failure at Deauville.

FORSTER, Edwin Morgan (1879–1970), 1901 travels for a year in Italy, Sicily, and Austria (*A Room with a View*, 1908; *Where Angels Fear to Tread*, 1905); 1903 cruise to Greece, returning through Florence; 1905 serves as a tutor for the family of Count von Arnim at Nassenheide in Germany, touring various towns along the Baltic on his way home (*Howards End*, 1910); 1912–13 to India with Goldsworthy Lowes Dickinson and R.C. Trevelyan; 1915–19 serves with the Red Cross as a "searcher" for missing soldiers in Alexandria (*Alexandria*, 1922; *Pharos and Parillion*, 1923); 1921 second trip to India, serving as secretary to Bapu Sahib, the Maharajah of Dewas State Senior (*A Passage to India*, 1924); 1935 addresses the International Congress of Writers in Paris, also visits Amsterdam with Bob Buckingham; 1947 and 1949 trips to America (1947 trip includes a tour of the American West); 1945 third trip to India; 1959 visits Italy; 1962 to Italy; 1963 to St. Remy, Paris, and Switzerland; 1964 again to St. Remy.

GISSING, George (1857–1903), 1876 travels to America to escape the disgrace of his imprisonment for theft (lives first in Boston, then Waltham, Massachusetts, then Chicago, then New York City); 1888–89 five-month tour of the Continent, including Paris, Naples, Rome, Florence, and Venice (*The Emancipated*, 1890); 1889–90 second Mediterranean trip, visiting Athens for a month and Naples for two months (*Sleeping Fires*, 1895); 1897 third visit to Italy, including stays in Calabria and Rome (*By the Ionian Sea*, 1901); 1898 on the way home, detours through Berlin; 1899 moves to Paris to live with Gabrielle Fleury; 1902 suffering from emphysema, moves to the south of France, near St. Jean-de-Luz, where he died the following year.

HAGGARD, Henry Rider (1856–1925), 1875 to South Africa as secretary to Sir Henry Bulwer, governor of Natal; 1876 visits Chief Pagate in northern Natal; 1877 in Pretoria to help raise the British flag in an annexation ceremony; 1880 visits South Africa (various romances, including *King Solomon's Mines*, 1885; *She*, 1887; *Nada the Lily*, 1892; *Marie*, 1912; *Child of Storm*, 1913; 1887 trip to Egypt (*Cleopatra*, 1889); 1888 travels to Iceland (*Eric Brighteyes*, 1891); 1890 travels to Mexico; 1900 travels to Italy and to the Middle East; 1905 to the United States to research Salvation Army settlements; 1912–17 travels around the world as a member of the Dominions Royal Commission; 1916 visits all Overseas Dominions to investigate prospects for "soldier settlements"; 1918 visits various parts of the Empire as a member of the Empire Settlement Committee.

HARDY, Thomas (1840–1928), 1876 with his wife, tours Holland and the Rhine, returning through Brussels where Hardy explores the field of Waterloo (*A Laodicean*, 1881); 1880 tours Boulogne, Amiens, and several towns in Normandy; 1882 lives for some weeks in Paris near the Left Bank; 1887 tours Italy ("Poems of Pilgrimage," 1902); 1888 another extended trip to Paris via Calais; 1893 to Dublin.

HARRIS, Frank (1855–1931), born James Thomas Harris; 1870–73 works in America; 1877–81 in Russia and Germany; 1881 travels in Greece, Turkey, Austria, Italy, and Ireland; 1891 (and subsequently) regular trips to the Riviera; 1898 to Africa; 1908 to America again (*The Bomb*, 1908); 1914 to France and then America; 1922 moves to Germany, then to Nice, where he dies.

HENTY, George Alfred (1832–1902), 1855 commissioned a lieutenant in the army purveyors' department, spent two months in Balaclava in Crimea; 1859 in Italy to organize the hospitals of the Italian Legion; as a war correspondent, he witnessed almost every major conflict of the period, including the Austro-Italian War and Garibaldi's Tyrolean campaign (1864); the Franco-Russian War (1870); the Second Ashanti War in West Africa (1874); the Turko-Serbian War (1876); and the North West Rebellion in Canada (1885). He wrote more than eighty essentially Imperialist adventure stories for boys, set in every continent, and at various times from the Roman invasion of Britain to the Boer War; some were based on his experiences as a war correspondent, though others derived from his journalistic and historical reading.

HEWLETT, Maurice Henry (1861–1923), from 1890 makes regular trips to southern Europe, particularly to Italy (*Little Novels of Italy*, 1899; *The Road in Tuscany*, 1904).

HOPKINS, Gerard Manley (1844–1889), 1860 travels with his father to southern Germany; 1867 to Paris; 1868 to Switzerland for a four-week walking holiday; 1883 visits Holland with his family; 1884 to Dublin as a fellow at the Royal University of Ireland, Chair of Greek at University College.

KIPLING, Rudyard (1865–1936), born in Bombay, India; 1871 sent to England to begin school; 1882 returns to India as newspaper writer (*Departmental Ditties*, 1886–90; *Plain Tales From the Hills*, 1888; *Soldiers Three*, 1888; *In Black and White*, 1888; *Under the Deodars*, 1888; *The Phantom 'Rickshaw*, 1888; *Barrack-Room Ballads*, 1892; *Many Inventions*, 1893; *The Jungle Books*, 1894–95; *Kim*, 1901); 1889 travels east from Calcutta, passing through Southeast Asia, China, and Japan (*Letters of Marque*, 1891) before arriving in San Francisco; toured Pacific Coast, Western Rockies, and Eastern United States (*American Notes*, 1891); 1890 travels to Italy and to America, and took a four-month voyage to South Africa, Australia, New Zealand, and India; 1892 on his honeymoon, embarks on what was to be an around-the-world tour, but halts in Japan due to the failure of his bank; 1892–96 lives in Vermont (*Captains Courageous*, 1897); 1898 to South Africa; 1899 to America; winters of 1900–1908 almost annual trips to South Africa (*Traffics and Discoveries*, 1904); 1899 serves as a war correspondent from South Africa during the Boer War; 1902 settles in Sussex, but continues to travel extensively.

KINGSLEY, Mary (1862–1900), 1893 travels alone to West Africa; 1894 travels to Old Calabar and the island of Corsica (*Travels in West Africa*, 1897; *West African Studies*, 1899); 1900 travels to South Africa, working as a nurse for the Boer prisoners; she died there after three months.

LAWRENCE, David Herbert (1885–1930), 1912 to Germany with Frieda Weekley; in August to Italy through Austria, settling at Gargnano where they lived until April 1913 (*Twilight in Italy*, 1916); 1913 trip on the Continent, including Germany, Switzerland, and Italy, where they lived at Fiascherino for eight months, then back to Germany (*Women in Love*, 1921); 1914 return to England, Lawrence traveling through France, where they stayed during the war; 1919 return to Italy, traveling through France; two months on Capri (*Sea and Sardinia*, 1921); 1922 leaves Europe, sailing south and east through Ceylon, where they spent six weeks, then to Australia (three months) (*Kangaroo*, 1923), then to San Francisco via New Zealand, Raratonga, and Tahiti; 1922–25 living near Taos, New Mexico, making several trips to Europe and Mexico (*St Mawr*, 1925); 1923 visits Mexico, then travels to New York through Texas and New Orleans; stays near New York for about a month, then travels back to California, spending a month in Los Angeles and visiting Santa Monica, Santa Barbara, and Palm Springs; returns to Mexico (Guadalajara), then sails for England in November; 1924 returns to the Continent, visiting Paris and Baden-Baden, then sails for New York in March, arriving in Taos the same month; stays in Taos for seven months; visits Arizona; departs for Mexico in October and stays in Mexico City for two weeks (*The Plumed Serpent*, 1926, *Mornings in Mexico*, 1927); 1925 returns to New Mexico, then to England in September; by the end of the year he and Frieda are at the Villa Bernaida in Spotorno; 1926 settle in the Villa Mirenda near Florence for the next two years (*Etruscan Places*, 1932); 1926 final visit to England; 1927 visits Germany; 1928 visits Switzerland; 1928 to the south of France, where he died.

MACKENZIE, Compton (1883–1972), 1900 summer in France; 1902 holiday in Spain and Northern Africa; 1903 holiday in France; 1913 purchases Casa Solitaria on Capri; 1914 to Dardanelles as a member of the Royal Marines; 1916 to Athens as a member of the Secret Service; 1917 on leave on Capri; 1920s living primarily on Capri; 1930 moves to Scotland, where he is elected rector of Glasgow University.

MAUGHAM, W. Somerset (1874–1965), born at British Embassy in Paris, where he lived the first ten years of his life; 1890 to Heidelberg to study; 1897 travels to Spain (first of a series a travels to gather material for writing) (*The Land of the Blessed Virgin*, 1905); 1914 as member of the British ambulance corps, serves five months at the front in Belgium and France; 1915 begins working for intelligence corps in Switzerland;

1917 travels to the United States and the South Sea Islands (Hawaii, Tahiti, and Samoa) (*The Trembling of a Leaf*, 1921); 1917 travels to Russia for British intelligence as part of a plan to prevent the Bolshevik takeover; 1922 and 1923 sails to Malaya, Borneo, the Pacific Islands, and Burma; 1933 travels to Spain; 1935 visits a penal colony in French Guiana; 1938 travels to India.

MEREDITH, George (1828–1909), 1842–44 attends the Moravian Brothers school in Neuwied on the Rhine (*The Ordeal of Richard Feverel*, 1859); 1849 honeymoon with Mary Nicholls on the Continent, including France and Germany; 1861 travels to Zurich; takes a walking tour of the Alps; visits Venice; 1862 travels to Tyrol (*The Amazing Marriage*, 1895); 1863 tours the French Alps with Lionel Robinson, crossing them and visiting Turin, the Lago Maggiore, Geneva, and Dijon; 1866 serves as a special war correspondent in Italy for the *Morning Post* (*Vittoria*, 1866); 1869 visits son Arthur at school at Hofwyl, near Berne, and takes him to Stuttgart (*The Adventures of Harry Richmond*, 1871); 1864–84 yearly visits to his second wife's brother's home in Normandy (*Beauchamp's Career, One of Our Conquerors*, 1891); 1865 in Venice (*Beauchamp's Career*, 1876); 1879 travels to France for health reasons, touring through the Auvergne to Nimes and the Riviera.

MORRIS, William (1834–96), 1855 travels to northern France to view churches and cathedrals; 1871 and 1873 visits Iceland (*Journals of Travels in Iceland*, 1913); 1858 again to northern France; 1896 sea voyage to Norway (in poor health).

OUIDA (Marie Louise de la Ramee) 1839–1908, 1850 to Boulogne with her parents; 1871 Continental travels to Belgium (*A Dog of Flanders*, 1872), Germany, Austria, Italy, and France, where she stayed for three months; 1874 until her death settles in Italy, first at a villa at Scandicci, near Florence; 1876 to Rome to gather details for the setting of *Ariadne* (1877); 1881 two-month visit to Rome; 1886 four-month visit to England; 1887 returns to Florence, where she moves out of her villa and into an apartment in Florence; 1894 moves to Lucca after her mother's death; 1903 moves to Viareggio in the same province, then to a villa at Camaiore; 1908 dies of pneumonia at Viareggio. Her other Italian novels include *A Village Commune* (1881) and *In Maremma* (1882).

RICHARDSON, Dorothy (1873–1957), 1891 to Hanover, Germany, as a pupil-teacher in Fraulein Lily Pabst's school, which became the setting for *Pointed Roofs* (1915); 1905–6 trips to Switzerland (*Oberland*, 1927).

SAKI (Hector Hugh Munro) (1870–1916), born in Burma; 1887 to France;

1889–90 to Germany, Austria-Hungary, and Switzerland; 1892 Switzerland again; 1902–6 correspondent with *Morning Post* in Balkans and Russia, witnessing the 1905 revolution (*Reginald in Russia*, 1909); 1906–7 correspondent in Paris; killed at the Somme.

STEVENSON, Robert Louis (1850–94), as a child, trips to the Continent for health reasons; 1878, *An Inland Voyage*, based on a canoe trip down the River Oise in France; 1879, *Travels with a Donkey in the Cevennes* about a walking trip through the Cevennes Mountains; 1876 summer spent at an artist's colony near Fontainebleau; 1879 to America in pursuit of Fanny Van de Grift Osbourne, traveling all the way across the continent to Fanny's home in Monterey, California (*The Amateur Emigrant*, 1895); 1880 honeymoon with Fanny at Silverado, an abandoned mining camp at Mount Saint Helena (*The Silverado Squatters*, 1883); 1880–81 prolonged expedition to Switzerland; 1882–84 living in Hyeres, France; 1887 second trip to America, first to Saranac Lake, New York; 1888 chartered yacht voyage to the Pacific Islands (Nuku Hiva, the Paumotus, Tahiti, Oahu, and the Hawaiian Islands); 1888–90 visits Australia and the Gilbert and Marshall Islands, finally settling on 400 acres on Samoa, where he died. *The Wrecker*, 1892; *Island Nights' Entertainments*, 1893; *The Ebb-Tide*, 1894; *In the South Seas*, 1896.

SYMONS, Arthur (1865–1945), 1889 first of many trips to the Continent—to Paris with Havelock Ellis; 1890 to Paris, where he meets Verlaine; 1891 visits Avignon, Arles, Barcelona, Madrid, Burgos, returning through Hendaye and Bordeaux; later that year to Berlin; 1892 to Paris; 1893 to Paris and Antwerp; 1894 visits Venice; 1895 visits Dieppe; 1896 visits Dieppe, Paris, and Ireland, the last with Yeats; 1896–97 visits Italy, with long stays in Rome, Naples, and Venice; 1897 visits Munich, Bayreuth, Warsaw, and Moscow; 1898–99 eight-month visit to the Continent, including France (Auvergne, Paris, Nimes, Avignon, Arles), and Spain (Toledo, Catalonia, Tarragona, Valencia, Alicante, Cordova, Seville, Don Juan, Figaro, Malaga, and Madrid); 1899 to Leipzig to examine portions of the manuscripts of Casanova's memoirs; 1902 long trip with his wife, which included Cologne, Munich, Bayreuth, Salzburg, Vienna, Budapest, Belgrade, Sofia, and Constantinople (*Cities*, 1903); 1903–4 to Switzerland, Italy (Bellagio, Venice, Ravenna, Riminia, Bergamo, Rome, and Pisa), and France (*Cities of Italy*, 1907); 1908 travels to Venice, Bologna, and Ferrara; he was confined to a mental institution near Bologna; 1924 to Paris and Burgundy with Havelock Ellis; 1925 visits Paris alone, then St. Malo, Pont-Aven, Nantes, Bordeaux, Toulouse, Albi, and Carcassone.

TROLLOPE, Anthony (1815–82), 1834 family moves to Bruges, Belgium, to escape bailiffs; 1835 returns to England; 1841 moves to Ireland as surveyor's clerk for the Post Office; 1858 visits Egypt and begins a long voyage to the West Indies; 1861–62 travels to America; 1868 again to America on a special mission for the Post Office; 1871 long trip to Australia to visit son (includes a trip to New Zealand); 1875 via the Mediterranean to Brindisi and Ceylon; later that year to Australia again, returning across America; 1877 to South Africa (*South Africa*, 1878); 1878 to Iceland (*How the Mastiffs Went to Ireland*, 1878); 1882 to Ireland to gather materials for *The Landleaguers*, 1883.

WALLACE, Edgar (1875–1932), 1896 to Simonstown, South Africa, as a member of the Medical Staff Corps of the British Army; 1899 stationed on the Orange River as Reuters correspondent in Boer War; 1900 sent to Beira, Rhodesia, for Reuters; after a brief return to England, back in South Africa as a war correspondent for the *Daily Mail*; 1902 in Johannesburg as editor of the *Rand Daily Mail*; 1903 to Canada as a reporter for *Daily Mail*; 1904 to Morocco, France, Spain, and Norway for *Daily Mail*; 1906–7 to the Belgian Congo to investigate rubber industry for *Daily Mail* (*Sanders of the River*, 1911, is the first collection of Sanders stories, a series that concluded with *Again Sanders* in 1928); 1913 holiday in Westende, Belgium; 1917 to Switzerland with Violet King, returning through Paris (these trips became annual holidays); 1919 to New York City to find an American agent; 1929 again to New York City and to Chicago (*On the Spot*); 1929 one of several visits to Germany; 1931 briefly visits Rome; later that year he traveled to Hollywood, California, on contract to RKO Studios, where he died.

WELLS, Herbert George (1866–1946), 1898 two-month trip to Italy; 1906 two-month trip to America; 1909 travels to France with Amber Reeves; 1914 visits Russia; 1920 visits newly formed Soviet Union to inspect Communism; 1924 moves to Grasse in the south of France, where he lived for three years, with occasional visits to his family in England; 1934 visits United States, where he interviews Franklin D. Roosevelt, and the Soviet Union, to interview Josef Stalin; 1939 voyage to Australia.

WILDE, Oscar (1854–1900), 1875 travels to Italy with his Greek tutor from Trinity College; 1877 travels to Athens and Mycenae, returning through Rome; 1881–82 spends a year in America and Canada (meets Longfellow, Oliver Wendell Holmes, Alcott, Whitman, and Jefferson Davis) (*Impressions of America*, 1906); 1883 travels to Paris (meets Hugo, Zola, Verlaine, Degas, Pissarro, and Toulouse-Lautrec); 1883 travels to New York City for opening of *Vera*; 1891 again in Paris; 1894 travels to Paris and Florence with "Bosie"; 1895 to Algiers with

"Bosie"; 1897 wanders through France and Italy after his release from prison; 1900 dies in Paris.

WOOLF, Leonard (1880–1969), 1904 to Ceylon as Cadet in the Foreign Service, disembarking first at Colombo, Western Province, then to Jaffna in the Northern Province; 1907 to Kandy in the Central Province; 1908 to Hambantota in the Southern Province as Assistant Government Agent (*The Village in the Jungle*, 1913); 1911 returns to England and marries (1912) Virginia Stephen (Woolf), *q.v.*

WOOLF, Virginia (née Stephen) (1882–1941), 1904 to Italy (Venice, Florence, Prato, Siena, and Genoa), with a week in Paris on return journey; 1905 trip to Spain and Portugal with her brother Adrian (Oporto, Lisbon, Granada, and Seville); 1906 visits Greece and Constantinople with her sister and brothers Vanessa, Thoby, and Adrian; 1907 to Paris with the Bells; 1908 to Italy with the Bells, including a week in Paris on the return trip; 1909 to France and Germany; 1911 travels to Broussa, Turkey, to help nurse Vanessa; 1912 honeymoon with Leonard—via Dieppe to Provence and Spain, then to Italy by sea, staying in Venice; 1927 with Leonard to Italy; later that year, visits friends at the Chateau d'Auppegard, near Dieppe; 1928 with Leonard to Dieppe, Cassis, and Fontcreuse; later that year to Paris, Saulieu, Vezelay, and Auxerre with Vita Sackville-West; 1929 to Berlin in January, then to Cassis and Fontcreuse in June; 1931 with Leonard, tours western France by car; 1932 with the Frys to Greece, via Paris and Venice, returning through Belgrade; 1933 car trip with Leonard through France and Italy; 1934 with Leonard to Ireland; 1935 with Leonard, car tour of Europe, including Holland, Germany, Italy, and returning home through France; 1937 with Leonard, car tour of southwest France; 1939 with Leonard, car tour of Normandy and Brittany.

INDEX

INDEX